I'm Sick, Ca

Pursuing healthy conversations in illness settings

By Mary I. Carson, Ed.D.

Copyright © 2010 by Mary Carson
(Standard Copyright License)

ISBN: 978-0-615-41783-7

*Dedicated to Irene, Theodore, and David
for their courageous illness journeys*

Introduction

This book serves as a "war on loneliness" whose defense relies upon ways to communicate when someone is ill. My volunteer experiences in hospice settings, combined with academic research, exposed the fact that no one is really helping patients and their loved ones *talk about* their experiences in ways that can lead to shared meaning, and ultimately, shared hope.

What happens when one leaves the doctor's office? How do patients, family, and friends *talk* about what the diagnosis means for each of them? As importantly, how do they talk about how the illness affects them over time? What should they say in the wake of seeming recovery or remission? What should they discuss in the wake of seeming failure?

Much has been written about doctor-patient communication, and many books and articles have been published counseling readers on how to think or feel in illness settings. But none of those resources examines the actual word choices that prompt those thoughts and feelings in the first place. Authors do not pay particular attention to the ways in which patients and caregivers weave words and phrases together in conversations that ultimately determine the levels of meaning they gain from their illness experiences.

Many patients and caregivers I have worked with reported they found themselves either at a total loss for words or completely frustrated at trying to say what they meant or felt. Yet, despite those challenges, most still yearned to be able to talk about their illness experiences together. They felt it was important to be able to do so, but they just didn't know how. As a result, both patients and their circles of support experienced distressing levels of loneliness and isolation from one another.

Sickness and pain are profound experiences for all involved, and as a result, words will never be able to fully explain the actual experiences. But words *can* help reduce negative thoughts and behaviors in light of those difficult experiences, and they can also create room for hope and shared meaning.

This book is designed to offer readers talking strategies that will empower patients and caregivers to enter into the courageous conversations illness settings demand. By learning how to better "say it like it is," readers will hopefully experience reduced stress and anxiety and find opportunities for personal growth and understanding that can be just as life-changing (in a positive way) as their original diagnoses.

Chapter 1

Saying It Like It Isn't

Communication may be regarded as a game in which the speaker and listener (writer and reader) struggle against the possibilities of confusion.
We must expect to be misunderstood.
We must expect to misunderstand.
We can try to minimize misunderstanding; we cannot hope to eliminate it.
--Ken Johnson[1]

To be human is to "report" to others. When change happens in our lives, we feel compelled to tell someone about that change. When new ideas occur to us, most of us want to share that information. When we experience new and different people, places, or things, we find ourselves wanting to talk about those experiences with others.

We rely upon spoken and written language to share our experiences with one another, and while I seem to be stating the obvious, it is that very obviousness that can lead to stress and frustration. We take for granted we are "saying what we mean" and "telling people like it is" by virtue of our word choices, but that assumption is rarely, if ever, true.

We forget words are symbols that have limited shared meanings and are somewhat arbitrary. Think of how language is created in the first place. As humans, we pull together sounds and alphabets that can be combined in various ways to create words. We then assign meanings to those words to help create order and agreement about use of those words. Finally, we put those words into dictionaries to clarify and record what those words will mean within our society, as well as how they should be used.

Once we put those shared meanings into place, we then *behave* toward those words accordingly. We physically and emotionally react to the words themselves based on the meaning we have invested in them. For example, if we see or hear the word "danger," most of us will at least give pause in some way and become more alert to our surroundings. On the other hand, the word "lullaby" prompts a more relaxed, almost blissful response. We can even "hear" what the lullaby sounds like in our minds.

For the most part, the shared meanings we develop help us function and interact with each other quite well. However, we forget words are also fraught with additional meanings that arise from how we personally use words over time. Our unique experiences contribute to the meanings of the words we use, resulting in our putting our "personal spin" on words, in addition to any dictionary definitions.

Examine this phrase: We should have gone to Norway. Does that mean anything special to you? I'm guessing not. But for me, the phrase has real significance based on a story I heard while growing up. Apparently the statement was muttered by a relative on her deathbed who mourned the fact she had not taken the time to visit Norway, a place she always imagined she would have loved to visit. Instead she had stayed at home to save money, just in case something happened.

Many people would suggest her statement was akin to "you only live once" or "stop and smell the roses," and those people would be right. But for me, the *more meaningful* statement is "We should have gone to Norway." It makes sense to me. It has meaning to me. But that's only because of my lived experience with that phrase.

These personal meanings are what prevent us from ever being fully understood by others, because as individuals, we can never know or experience fully what another individual knows or experiences. Words and phrases cannot actually replicate our experience; they can only stand for the experiences we are trying to describe. As a result, we are all forced to "interpret" the words we use when we talk with one another, which can lead to confusion and misunderstanding.

Let's say I reported to you "I bought a new laptop." Chances are you envision my having purchased some portable technology on which I can create documents, go online, send emails, etc. However, the type of laptop you picture in your mind may be very different from the laptop I actually bought. You would either imagine a laptop *you* would buy yourself, or you would imagine a style of laptop you think I would purchase based on what you know of my personality, sense of style, career, etc. Until and unless I clarify further what my new laptop is like or you actually see the laptop itself, you will have a limited understanding of my new laptop. Yet you'll likely say something like "that's great" or "how wonderful," and we'll both walk away from that conversation thinking we fully understood each other.

While this is a simplistic example, it shows the ways in which we are able to function at least on a surface level in sharing meaning with one another on a daily basis. Even if you did not fully understand what my laptop actually looked like or was fully capable of doing, neither of us was stressed or upset because we had a basic understanding about the form and function of the laptop itself. We both felt confident in our shared, albeit limited, meaning.

We begin to have problems, though, when we try to report more abstract concepts—ambiguous words and phrases that have huge personal meanings. For example, what does "love" mean? What does "justice" look like? What is "patience?"

Because this book relates to words used in illness settings, take a moment to think about the word "illness." What *is* "illness" exactly? What is a "life-limiting diagnosis?" What is a "chronic disease?"

Similar to my laptop example above, if I were to report to you "I learned I have cancer," you will likely ascribe a meaning to my cancer that is based upon your own experiences with cancer. Perhaps you had a grandfather die from complications of lung cancer, yet you also know a woman who survived breast cancer. Maybe you, yourself, have had cancer. Any or all of those prior experiences will necessarily shape the meaning you assign to *my* cancer. But there is no way my cancer can be exactly the same as others' or your own. Still, after hearing my report, you will assume you essentially know what I am talking about.

I, too, as the reporter of that information would go

through a similar process. I would assign meaning to my cancer in light of what I knew of others' cancer or what society had told me about cancer, especially if I had never experienced it before. Until I gained more information about my type of cancer, I would be talking to myself and others at a surface level of understanding that would likely not be positive or accurate. However, like you, I would assume I knew what I was talking about.

As importantly, we would both also *behave* in certain ways toward my illness that reflect the way we talked about it. For example, if most of my language choices in describing my illness were highly anxious, we would physically and emotionally respond to those words in ways that reflect my anxiety. Perhaps we'll cry, talk softly, and experience a heaviness in our hearts. On the other hand, if I use more positive language choices, we might then smile together, hold our heads up, even go out to dinner and celebrate our friendship.

The overarching challenge we face when talking about illness is the words we use are rarely up to the task of helping us say it like it is. We strive to paint word pictures that fully explain what we are feeling and experiencing, but that is just not possible. Inevitably our words will fall short. This is where stress, frustration, disappointment, and numerous other negative feelings emerge when talking to others — and ourselves — in illness settings.

We can also find it difficult to talk with others about an illness experience over time. The words and phrases we relied upon at the time of diagnosis often prove to be inaccurate or inadequate in representing changes over time. Also, the word pictures we created to predict the future often do not match what we actually experience along the journey, and as a result, we find ourselves struggling to say exactly what we feel or mean.

So in light of all the challenges described above, should we simply not talk to one another when we are sick?

Absolutely not. It is important to talk with others when we are sick or hurting. It is what makes us human, it is what allows us to develop and maintain our support networks, and perhaps most importantly, it is what helps us create meaning about our experiences.

We do not need to stop talking; we instead need to learn strategies that lead to healthy conversations when we are sick.

There is a body of communication theory and strategies referred to as "General Semantics" (GS) that can help us respond to the challenges we face when talking to one another in illness settings. GS recognizes the symbolic nature of language and offers ways we can be more mindful of our language choices to better insure shared meaning. GS also focuses on the relationship between what we say and how we behave or respond to what we say.

The founder of GS, Alfred Korzybski[2], was a Polish engineer who wanted to help people learn to better say it like it is to help reduce anxiety and stress that can be prompted simply from talking to one another. Korzybski recognized that what we say is how we think and behave, so we needed strategies for aligning our words and phrases more closely with reality.

Specifically, Korzybski posed three key premises that serve as the foundation for this book:
1. The word is not the thing
2. The map is not the territory
3. The map reflects the map maker

The Word is Not the Thing

Pretend my friends and I collectively decide to rename our TVs "socks." If I then later reported "the picture on my socks seems a bit fuzzy," those of us who understand the new meaning for socks won't be surprised by my comment. People may even offer suggestions on how to fix the problem with my socks. As long as there is agreement about the meaning of a word (symbol), life can move forward.

Still, even though I might succeed in having everyone agree to call TVs socks, think of how difficult and even uncomfortable that might be for us to do naturally. After all, we know TVs are *not* socks. TVs *are* TVs, and socks *are* socks. Each of those words bears a definition we have agreed upon for a far longer period of time, so my new definition literally feels uncomfortable.

Our existing responses to the words "TV" and "socks" can also prevent us from thinking about those objects in new ways. We have been told TVs *are* electronic devices on which we watch shows. Socks *are* cloth objects we put on our feet. Could you imagine, though, there are TVs in this world that do more than just show TV shows or socks that are not made of cloth? Certainly, yet our language does not easily adjust itself to those differences.

Our knee-jerk reactions are to cling to the words "TV" and "socks" in the singular format we originally assigned and learned. Doing so makes life easier to categorize and predict, and we also think it helps us communicate since we're relying on what we have already agreed is correct and appropriate. We both learn and believe the word *is* the thing.

Now consider the words "cancer" or "biopsy." It is highly unlikely I could get my family and friends to comfortably agree to replace the word "cancer" with the phrase "a minor setback." We are simply too programmed in believing cancer is always bad, and therefore, interchanging the words "cancer" and "minor setback" would make us uncomfortable on many levels. The same could be said if I tried to replace the word "biopsy" with the phrase "routine examination." A biopsy may be routine for a doctor performing the test, however, the rest of us would rarely consider the test to be routine.

Highly-charged illness-related words prompt a visceral reaction that not only relates to how we *think* about the words, but also how we actually *feel* about and *behave* in response to the words. Think of what you physically felt when you first read my ideas for exchanging words, as well as your mental challenge in doing so. Those reactions demonstrate how the meanings we assign to words—which, again, are really just symbols—literally affect our feelings and behaviors toward others and our environment. Words have power over us.

To help us break free from the limited realities we impose upon ourselves through collectively assigned meanings, we must remember the symbolic nature of language and work harder at really saying what we mean. Chapter Two offers strategies on how to do so.

The Map is Not the Territory

Whether we use a street map, a topographic map, or a newer GPS system, maps are intended to help us get from point to point. Maps are also designed to describe specific territories in terms of geography, road systems, towns and cities, etc. They help us know what a specific area or territory is "all about."

Imagine, then, you want to venture from Pittsburgh to Philadelphia, but you only have a map of Wisconsin. How helpful would that Wisconsin map be? Not very. What if I were to suggest you should at least *try* using the Wisconsin map? Chances are you would become frustrated very quickly. The problem? The map simply does not reflect the territory through which you want to travel.

We create and rely upon "verbal maps" to navigate our personal lives in much the same way we use traditional geographic maps. We weave words and phrases together to describe "territories" of "how things are," and then we travel within the verbal realities we create for ourselves.

To better understand this concept, first think of all you "know" at this point in your life. All you have learned in various educational settings, all you have seen on TV, all you have read in the newspaper, etc. It is likely your brain holds quite a wealth of information.

Now consider how much of your knowledge of "how things are" comes from first-hand experience as compared to information you have heard or read about from others. For most of us, the knowledge we gain from first-hand experience is dwarfed by that which we have learned from other sources ("reports," "reports of reports," "reports of reports of reports," etc.).

The challenge of relying upon reports of others is that the likelihood for confusion or misinformation rises in direct relationship to the number of people sharing the reports. Think of the children's game "Pass the Secret" in which one person whispers a message into her partner's ear, and then the message is whispered to that partner's partner, etc. The fun of the game is that the original message becomes mangled and distorted as it is passed from person to person. By the time the message is passed among five or more children, it barely resembles the original.

A similar phenomenon happens as we hear or read the reports of others and then combine them into "maps" that help us navigate our lives (the territory). Like the children's game, depending upon how far away the reports are from their original sources, we may learn or inherit faulty reports. We all know time changes things, including the reports upon which we rely, but for some reason, we find it easier to cling to our original reports. Even when we are openly confronted with new or different information, many of us choose to ignore that information. Instead, we move forward with our limited maps which ultimately inform our attitudes, beliefs, and values.

Unlike the children's game, though, there is no fun in

discovering our maps *are* distorted or otherwise faulty.

An example relates to a report many of us heard while growing up: "hard work pays off." This verbal map of what happens in life has been reported for decades, and for many, that map was (and is) accurate. Those especially who held factory jobs in the 1940's and 1950's worked very hard (long hours and lots of physical labor), and many of those workers indeed succeeded in life. They retired comfortably, helped their kids, and enjoyed their grandchildren.

But as technology advances reshaped shop floors, "hard work" did not necessarily pay off. In fact, hard labor soon became automated and people found themselves out of a job. Those workers were stunned that what had been true for their parents' and grandparents' generations was no longer the case. The verbal map of hard labor paying off was no longer accurate, and the stress of trying to force that map upon the changing situation grew in direct proportion to the pace at which technological change happened.

Examples of other verbal maps many of us have been exposed to include "you need a college degree to get a good job," "politicians should expect their private lives to become public," "you're either with us or against us," and "don't swim for an hour after you eat." Certainly we can imagine, and have likely witnessed ourselves, instances in which these maps are untrue. Yet our knee-jerk response is to accept them because they ring so familiar.

We use verbal maps like these to describe and predict how life "is." Not all of our maps are incorrect or bad for us, and when they do accurately reflect the territory we experience, our stress is reduced. However, when our actual experiences no longer match our verbal maps, stress and anxiety can increase.

We need to develop new verbal maps.

In terms of illness settings, a common verbal map we hear is "eat right, exercise, get plenty of rest, and you'll live a long life." That map may work for many people, but what about the man who does just that for 50 years, only to find out on his 51st birthday he has been diagnosed with Lou Gehrig's disease?

"Smoking leads to lung cancer" is another map we hear, and while smoking has been proven to be related to lung cancer, can you imagine the surprise of the 62-year-old woman who never smoked a cigarette in her life finding out *she* had lung cancer?

In order to have healthy conversations with others, we need to be mindful of the maps we bring to those conversations, especially when facing a serious illness. A scary diagnosis can turn our verbal maps upside down. To that end, Chapter Three will offer ways for us to safely renegotiate the maps that seem to no longer fit, as well as offer strategies for creating new maps in which we can find comfort.

The Map Reflects the Mapmaker

We cannot escape ourselves when creating the verbal maps that drive our thoughts and behaviors. All of the maps we create are reflections of our biological and neurological selves that we simply cannot avoid. Our brains and our five senses play especially critical roles in shaping how we perceive our environment, and as a result, they shape the conversations we have with others and ourselves. Just as we talk in ways that reflect our ability to *think*, we also talk in ways that reflect our ability to *sense* our environments.

All of the reports we share with others and ourselves reflect both the strengths and limitations of our nervous system — our brain and the neurological network housed in our bodies. In many ways, our brain and nervous system allow us to perceive and describe our environments in fantastic ways. But we must also remember our brains and senses simply cannot process everything that happens around us. Some stimuli in our environment are simply beyond our ability to actually perceive. For example, we cannot see or touch microwaves or radio signals. We cannot hear sounds above a certain decibel level.

Even when we are able to sense stimuli in our environment, we still have difficulty in that we are bombarded by such an abundance of stimuli, there is no way we can take it all in at once. Anyone who has gone to a circus performance will relate to this concept. Trying to see and hear everything going on in three separate rings can be impossible, so we are instead forced to attend to some things and miss out on others. As a result, our maps feature some of the landscape, but not all of it.

Choosing where to focus our attention (or not) further influences the verbal maps we create. My choice to solely focus upon the trapeze artist means I cannot also watch the lion tamer. Does that mean the lion tamer does not exist? Obviously not, but any attempt on my part to describe the lion tamer's performance will necessarily be diminished in contrast to the trapeze artist. Ultimately, all the choices I make in what to observe at the circus will lead to a verbal map of that experience that is unique to me. Meanwhile, other audience members will make their own choices too, and so their verbal maps will likely be different from mine.

So whose map is most accurate? None of ours, as we were all limited in our ability to attend to the whole event.

Finally, our "perspective" on a matter also influences any map we might create. The position from which we speak—literally and figuratively—will necessarily shape any report we make about any situation. Our ability (and more often our inability) to view situations from different perspectives necessarily edits the amount of information we can glean from our surroundings, which, in turn, leads to verbal maps that tell only part of the whole story.

Take a moment and look at your immediate

surroundings. If you are reading in a room, can you see a window or a door? Can you see what is outside that window or door? What happens when you move to another space in the room? What can you see that you could not before? As mapmakers, where we position ourselves influences the verbal maps we create for ourselves and others.

We cannot hope to escape ourselves when it comes to creating our maps. "Objectivity" is impossible, and we can only craft reports that are relevant to ourselves and are necessarily incomplete because of our neurological limitations.

This conclusion is not meant to make us feel inadequate or suggest we shouldn't even try to engage in the mapmaking process. Instead, it is meant to remind us the world is filled with infinite possibilities and yet-to-be-discovered "aha's" that we can fold into our maps along the way to make them more meaningful and hopeful.

Moving forward from here...

The following chapters build upon each of these three premises by offering specific strategies that can help patients and caregivers pursue healthy conversations in the wake of an illness. We are what we say, and in that regard, words matter. Learning how to make conscious language choices when talking to ourselves and others not only enables us to speak more accurately, but it also helps us create room for hope and opportunities for personal growth when facing serious medical issues.

It is important to remember these strategies can help everyone involved in a conversation about illness—not just the patient. While many examples do focus upon the patient, suggestions for caregivers and other members' of a patient's circle of support are also offered.

Chapter 2

The Word is Not the Thing

We need new words for illness, not to be drawn from the lexicon of complaint or from the book of invalidating phrases…We need subtle words that allow for degrees between healthy and sick, descriptive words for naming the in-between states where we spend much of our lives. We observe these states with silence, or catchalls—"I'm not feeling well"—but our lives are more complex. They ask not for confession, but for calibration, so that we may tell one another how we are.
--Janet Sternburg[1]

It is not only true that the language we use puts words in our mouths; it also puts notions in our heads.
--Wendell Johnson[2]

"There's no easy way to say this, but my husband, Jim, has cancer. We just found out this week. The tests confirmed our deepest fears."

"Cancer?!? I'm so sorry for you two!"

"He's feeling really low, wondering what he did wrong to get cancer. You know what a health nut he is."

"Cancer. I cannot believe Jim has cancer."

This brief conversation demonstrates how easily stress and anxiety can develop when we believe the words we use are actually the things they represent.

Notice how both speakers use the blanket term "cancer" as if it is enough to completely describe Jim's unique diagnosis. Herein lays the problem. Neither speaker clarifies what Jim's *specific* cancer entails biologically or experientially. No mention is made as to the type of cancer, the location of the cancer, the possibility for beating the cancer, etc. Instead, the two friends solely rely upon the word "cancer" as a single definitive concept.

Furthermore, each speaker uses the word "cancer" in ways that reflect her own personal understanding of that word. If either speaker has experienced cancer herself, she will likely impose the meanings she gained from her own experience with cancer onto Jim's experience. If neither has directly experienced "cancer," then each will impose what she learned about cancer from others' experiences. And if any of those prior experiences were largely negative or positive, the speakers will believe Jim's experience will also be largely negative or positive.

The speakers forget there is no way Jim's experience with cancer can be like any other person's experience with cancer. Even if he has a similar diagnosis, how the illness affects his body, mind, and soul simply cannot be identical to any other person's experience. Therefore, the word "cancer" is not the "thing" Jim has, because the small two-syllable word cannot fully represent the infinite diversity of illness experiences that can result from a cancer diagnosis.

To make matters worse, each speaker's reliance on her personal spin of meaning for the word "cancer" will influence her behavior toward both Jim and Jim's illness. They will behave toward Jim in direct reflection of what they say about Jim.

These challenges demonstrate why healthy conversations in illness settings require us to remember the words we use are not the things they represent. Easier said than done, but there are a number of strategies we can pursue to help us avoid the pitfalls of personal meanings and assumptions.

Words matter.

Despite the fact words are limited by their symbolic nature, we still need them to help us converse with one another. The good news is we've got plenty of words to choose from in our respective languages. The challenge, though, rests with our willingness to purposely vary our word choices to best say what we mean.

Think of the number of words housed in a dictionary of the language you speak most fluently. What percentage of those words do you use on a regular basis? I'd bet most of us use a very small fraction. Certainly we're still able to have meaningful conversations with others, but at the same time, our limited word sets can limit our ability to fully explain what we think and feel. This is especially true when trying to communicate more abstract thoughts and emotions.

For example, think of the differences among the words "disappointed," "frustrated," and "angry." What do these words *mean* to you? How do they differ exactly? What examples might you offer to demonstrate those differences?

For me, being "angry" conjures a more heightened sense of emotion than does being "disappointed." Angry looks like someone yelling, maybe kicking something or shaking a fist, getting red in the face, etc. On the other hand, "disappointed" suggests a level of sadness to me. Disappointed makes me think of someone's shoulders slouching, the wind knocked out of them, a sense of sadness in not getting what was truly, truly desired. I also see "frustration" as easily leading to "anger" but still not quite there. "Frustration" is like forgetting something you should normally know or not being able to finish a puzzle. For me, it is kind of a nagging feeling.

Your personal spins on these words might be similar to mine, but there is also a good chance we would differ to some extent. Perhaps your meaning of "angry" does not mean loud and in-your-face behaviors. Maybe it comes about through avoidance of others or otherwise refusing to participate. If that's true, you can see why I would be confused if someone told me you were angry, as I did not witness any tantrums on your part. I might have actually thought you didn't care!

Despite these differences of interpretation, we can still enhance understanding in our conversations by expanding our word choices and then carefully choosing the best words possible. Larger vocabularies can help us better describe what we think or feel. That does not mean we have to learn "big words" or totally avoid words we have been using over time. But having access to a large portfolio of descriptors can help increase shared meaning and reduce misunderstanding.

Take a moment and think of the different ways you could report you were "happy." Push yourself…take at least a couple minutes. Then look at my list below:

blissful	charmed	cheerful	contented
delighted	ecstatic	elated	exultant
fantastic	giddy	glad	gratified
high	joyous	jubilant	merry
pleased	satisfied	thrilled	tickled

Notice how each word has a different "spin" on being happy that makes it slightly different from the rest. Certainly what you might view as "delighted," I may view as "elated," so we might still need to clarify our ideas a bit as we talk. But the fact we worked hard at choosing the right word from the start can help reduce that effort.

Illness settings especially require us to work harder at saying what we mean. The word "sick," for example, is fairly ambiguous in that it could describe a myriad of physical and emotional states of being. Again, try to imagine different words you might use instead that would better describe the specific aspects of feeling sick. Then look at my list:

ailing	sore	feverish	queasy
weak	cold	dizzy	nauseous
exhausted	achy	stuffy	pressured
throbbing	short of breath		

I have found using a thesaurus can be very helpful for identifying new and different words. Be assured, my goal is not to dazzle or stump others with words one might not normally use or understand. But there have been times when I experienced genuine relief after discovering just the right word(s) to express what I wanted to say.

Bland reports like "I'm feeling so-so today" or "I think I'm better" do not help listeners fully understand a patient's experience, and as a result, listeners may respond inappropriately. It's hard to say or do the right thing when one has limited information. And talking about pain in illness settings can be especially challenging for both patients and their circles of support (see Chapter 6).

Still, the harder we work to better say what we mean, the more we can avoid misunderstanding and hurt feelings.

To that end, below are examples of two "surface statements" I often hear in illness settings. Each surface statement is followed by "reframed statements" that replace the generic terms "better" and "bad" with more specific word choices. Note how the reframed statements paint a clearer picture as to *how and why* the speaker is feeling better or worse which, in turn, enables listeners to better understand the patient's experience and respond appropriately.

Surface statement

"I'm better."

Reframed statements

"I'm energized from my nap."
"I feel stronger since taking that aqua aerobics class."
"I am more confident after hearing the test results."

Surface statement

"I'm feeling bad."

Reframed statements

"I'm feeling miserable from the chemo treatment."
"I'm devastated I have to miss the family gathering tonight."

> "I'm stunned at how many pills I have to take each day."

We also need to recognize the contexts within which we use our words. The saying "consider the source" means a great deal in this regard.

Our word choices are influenced by our interactions with our environment over time, and as a result, the meanings we assign to the words we use are constantly changing.

Slang is a good example of this phenomenon. "Hip" words and phrases often emerge from societal influences, especially the music and entertainment industry. Think of the words you used to describe something as "cool" when you were in high school and compare it to the terms kids in high schools use today. If you're like me, they hardly match. The context has changed, the culture has changed, societal values have changed, and therefore, the language choices and their *meanings* have changed.

Now consider the word "priority." Think of how its meaning has changed for you over time, especially in response to personal life events such as pursuing a degree, landing a good job, getting married, having children, wrestling with health issues, etc.

Think also how your use of the word "priority" has evolved in light of the political and economic changes you have witnessed. I would bet some of what you considered to be a "priority" 10 years ago does not carry the same sense of urgency today.

Before I became a mom, I used to think it was a priority to make sure dishes were cleaned and stowed at night. After I became a mom, it was more important to help get homework completed and lunches packed for the next day. The dishes could wait—they were no longer a priority. Changing circumstances in my life resulted in changed meanings for the words I had used.

I suppose someone could consider my change in meaning as somehow being fickle, especially if that person still viewed an empty sink at night as being a priority. But the fact is the meanings we assign to the words we use directly relate to the experiences within which we use those words...the *context* behind what we say...and that's okay.

Below are four identical statements followed by a context within which they might be spoken. As you read through the list, be mindful of your own reaction as to how "important" the choices are in regard to each context.

> *This is an important choice.* (Choosing the type of cake to serve at a wedding.)
> *This is an important choice.* (Choosing a home to purchase.)
> *This is an important choice.* (Choosing a chemotherapy regimen.)
> *This is an important choice.* (Choosing a camera to take on a vacation.)

You likely experienced some sort of judgment about the actual importance of each statement in light of the proposed contexts and your own experiences. Certainly choosing a camera is a very different experience from choosing a chemotherapy regimen, not to mention purchasing a home. But do those differences make any of the statements more "correct" in terms of using the word "important?" That depends upon the person speaking.

My knee-jerk reaction as the listener is the chemotherapy regimen is most important, as it could be a life or death decision. But that is a judgment on my part that may or may not be accurate for the speaker.

The speaker may instead believe the choice of camera is most important because she is concerned about capturing the best pictures possible on her vacation. Since her husband has been diagnosed with a serious illness, this may be the last vacation they have together.

Which of us is right? There is no right or wrong. There is simply *perception,* which creates a great deal of room for misunderstanding. So we need to be mindful of the contexts within which people are speaking and then try to resist jumping to assumptions and judgments about what they say.

Restating or "reframing" one's ideas is not an easy task. One may literally have to pause and think for a time before speaking. Most of us are not used to being so careful about what we say, relying upon the fact friends and family members know what we really mean. But as noted above, that assumption is faulty, and talking at surface levels cannot lead to full understanding, let alone creative solutions.

Asking ourselves the following questions can help us say more closely what we mean:

- What exactly am I trying to describe? A physical sensation? An emotional sensation?
- What are my senses telling me? What do I see, hear, smell, taste, touch, etc.?
- What other words might I use to describe what I am sensing? Is the word I chose first the most appropriate? Accurate? Should I combine words?
- Do my words take into account the way my sensations change over time? Should I add new words? Eliminate old words?
- What is the context within which we are having this conversation? What is going on around me? Around us? How is that influencing what I say?

Words cannot say it all.

Even if we work hard at stringing words together that have the closest meaning possible to what we are trying to describe, some information will always be left out. We can never *fully* describe a person, place, thing, idea, etc. no matter how many words or phrases we use, because words are symbolic in nature. We use words to *stand for* ideas and experiences; they cannot recreate actual experience.

Think of times when you have been introduced to others and they ask the familiar question "what do you do for a living?" Assuming the person who asked does not have hours to listen in full detail about your job responsibilities, you likely summarize what you perceive to be the most important aspects of "what you do." The choices you make in what you share (and do not share) will necessarily influence how both you and the listener interpret and react to your job.

If you focus only upon the frustrations of your current post, you both will sense the job as a largely negative experience. On the other hand, if you talk about the benefits you gain and creative opportunities you enjoy, you both sense the job is largely positive. Which interpretation is *correct?* Neither, since each does not fully describe the actual phenomenon of your total work experience.

Furthermore, you likely used single words or short phrases to describe your job duties for the sake of convenience and time. For example, you might have used words such as "challenging" or "stimulating," or phrases such as "puts food on the table" or "better than my last job."

Again, while these words and phrases do describe aspects of your job, they still do not explain *all* there is to what you do. They are limited, subjective, and incomplete descriptions that only scratch at the surface of what your lived work experience is really all about. And unfortunately, those limited and subjective words and phrases result in limited and subjective responses on both your parts.

Limited information leads to limited responses, so to better explain ourselves and paint more accurate verbal pictures, we need to adopt some key strategies.

First, we must remember no single word or phrase can completely describe the entirety of what the word or phrase is supposed to represent. Easier said than done!

Look again at the conversation at the beginning of this chapter about Jim's "cancer" to see what happens when this concept is forgotten. The speakers use the single word "cancer" to completely describe Jim's diagnosis, and ultimately, Jim as a person. They have forgotten how limited that two-syllable word really is for describing Jim's experience, and as a result, stress and anxiety are given enormous room to grow. The word cancer simply cannot fully explain *all* aspects of the situation.

Compare the following two reports:

(1) "Jim has cancer."
(2) "Jim has a 2-centimeter cancerous tumor located in his upper left lung today."

The first report forces any and all interpretations of Jim's diagnosis to rest *solely* upon the single word "cancer." It is as if Jim has been draped head to toe with cancer, and he himself almost disappears. Jim essentially *turns into* cancer, and we lose sight of Jim the person because we become so focused on the negative word. As a result, we grow anxious and easily think of the worst that could happen to Jim.

While the second report could also be interpreted as overwhelming and frightening, it more accurately describes the scope and timeframe of Jim's diagnosis, which leads to opportunities for more varied and thoughtful responses. The fact the tumor is currently located in Jim's upper left lung reminds us there are other body parts that are *not* cancerous. Describing the size of the tumor also helps us better imagine the cancer as being a very small part of Jim. We do not think of Jim as having cancer head to toe. Using the word "today" also forces us to remember things can and do change over time.

Certainly that small tumor could wreak havoc on Jim's life, but our reaction to the second report will likely be more positive (and hopeful) because it offers a more complete verbal map that allows room for other aspects of Jim to appear to us at the same time as the illness. In addition, the second report can lead to a more open mindset to possible plans for treatment and accommodation.

Think about the word "normal"—a word often featured in illness settings. Patients and their circles of support often report their desire to go back to a time when things were more "normal," and their wistful yearnings suggest that normal is always "better." But what exactly *is* normal, especially in light of the fact time changes everything?

More often than not, patients tend to think of their healthy experiences prior to their diagnoses as having been "normal," and illness or injury experiences as having been "not normal." For many of those patients, the number of healthy experiences outnumbered the number of illness or injury experiences, so good health was deemed to be normal.

Yet, those same patients also likely experienced enough negative life events to know it was also normal for bad things to happen to good people, including themselves. If pressed, it is likely none of them would suggest there was ever a time they were totally immune from sickness or injury, and they might even suggest setbacks in life were inevitable—were normal.

So which version of normal is *correct*? Neither, since the word normal is so severely limited in describing one's total lived experiences. The word normal cannot fully describe *all* of an experience. That is why when we hang our hats on such a limited and inaccurate interpretation of the word normal, we grow frustrated and disheartened when we discover we cannot achieve that which was inaccurate in the first place.

The medical arena also tends to use the word normal as it were something in and of itself. Doctors interpret test results, pains, side effects, etc. as being "normal or not."

Certainly medical practitioners' use of the word rests with observations of repeated *similar* experiences among large samples of patients, however, we all know there are always exceptions to the rule.

So is the person who is the exception to the rule not normal? Might she in fact be normal in terms of *her* subjective experience?

Compare the following two reports:

(1) Your test results were not *normal*.
(2) Your test results showed an increase in _____, which is a higher number than the last time we ran the test.

The first report bears an almost judgmental tone, since we tend to think of the word normal as being something good. Who wants to be told they are not normal in any shape or form?

Our society focuses upon consistency and conformity more than we realize, so the word normal bears some heavy baggage. As a result, the patient being told her test results were not normal will easily experience stress and discomfort at being told she is somehow different from what is expected. After all, they are *her* test results.

The second report still speaks of the patient's

difference in terms of results, however, it does so in terms of *a previous test* rather than in general terms.

The second report also focuses on what exactly was different and simply describes the difference as being "a higher number." By avoiding the word normal, judgment about the *person* is removed.

Certainly the patient can become anxious about the test result showing a difference, but hopefully, the patient does not view her entire being as out of sync with what should be. By focusing on the difference in specific terms, the patient can pursue conversations that examine possible next steps without her wondering if she is somehow lesser a person for having experienced the test results in the first place.

Many words we use to describe our daily lived experiences are necessarily subjective, situational, and *incomplete*. Therefore, we need to push ourselves to use more words to paint more complete verbal maps that help lessen or avoid the judgments, biases, and stereotypes bound up in single words or short phrases.

One technique we can use to demonstrate the limited nature of what we are saying is to simply add the words "et cetera" or "etc." to what we say or write. When we use the word et cetera at the end of a sentence, we remind ourselves and others that there is more to say on a matter.

Use of et cetera also simply gives us a break from having to identify all that we mean when we could literally go on and on but are too tired to do so. "Etc." allows for the fact there is likely more to know, something has been overlooked, *etc.*

In an illness setting, use of the word "etc." helps speakers let listeners know "there is more to be said...I just cannot say it all." A huge frustration among seriously ill patients is their inability to fully explain all they are experiencing. Caregivers are equally frustrated trying to fully understand the patient's experience, so they continue to ask questions for clarification, fueling the patient's frustration.

Relying upon the word "etc." can help remove the stress of trying to explain something that is literally beyond words.

"How are you doing?" in an illness setting is as regular—and frustrating—a question as the "what do you do for a living?" question noted above. There is no way a full report can be crafted by patients facing life-limiting illnesses. There are simply too many physical, emotional, and even spiritual dimensions that defy accurate description.

Any effort to fully explain will not only fail, but will also frustrate, disappoint, upset, etc. the patient. Use of "etc." helps remind both the speaker and listener of that fact.

Use of "etc." in reporting a negative situation

"When I'm sitting in the chair, I think about all the things I'm not getting done: the laundry, the grocery shopping, cleaning, etc. It is so damn frustrating."

Use of "etc. in reporting a positive situation

"The outpouring of support I'm receiving makes me feel humbled, loved, valued, etc. It has really made a difference in our family's lives."

Caregivers and medical personnel can also use "etc." when trying to describe their own experiences.

Use of "etc." in reporting a negative situation

"She has some really tough days when she just thinks about questions like why me, what will happen now, will I lose my job, etc."

Use of "etc." in reporting a positive situation

"We're really sorting ourselves out in regard to what he can or cannot do...climbing stairs, using the bathroom on his own, getting dressed, etc."

Again, the goal for using "etc." is to reduce the frustration of trying to say it all. It grants permission to leave the door open to other possibilities.

We have to work harder at making appropriate word choices to reflect our reality. No single word or phrase can accurately describe the fullness of anything, so we must instead strive to weave words and phrases together to create more complete verbal maps. In that regard, key questions to ask ourselves include:

- Am I saying it all as best I can? IS that all?
- What else could I add to what I am saying?
- Am I relying upon a single word to describe something I sense or think? Is that word accurate and complete?
- What is missing from the word I am using to describe something I sense or think? What more could/should be said?

"Abstracting" ourselves from miscommunication

Another challenge we face when talking with one another is we tend to use our words as labels that help us categorize people, places, ideas, etc. to make our conversations more convenient. We make words stand for a wide variety of concepts that we think share similarities so we don't have to work so hard at what we say.

Think, for example, about the word "snacks." What *are* snacks? Potato chips? Peanuts? Pretzels? Corn chips?

Each of these foods is quite different from the other, however, we tend to lump them together and refer to them as "snacks." So if I tell you I'm bringing snacks to your party, how surprised would you be if I brought popcorn? Likely not too surprised, as most of us consider popcorn to be a snack. But what if I brought pickled herring?

When we only consider similarities in choosing our words versus looking for differences, we end up doing what is called "abstracting." We move away from the individuality of people, places, or ideas and "abstract" them in ways that ignore the differences and focus only upon the similarities.

Think again about my variety of snack options. Would you say they are more similar or more different? It seems to me the only real similarity among them is how or when we eat them: as snacks. In and of themselves, though, they really are quite different in terms of flavor, food source, ingredients, etc. Yet we find it *easier* to simply call them all snacks.

Here is another example. If I told you I purchased a new chair, even though you did not see my new chair, you would assume I would use it to sit upon (and of course, you'd be right). Your knee-jerk assumption comes from your "abstracting" the word "chair" at a fairly high level: the *function* of the chair.

You did not concern yourself with the design of the

chair, the materials from which the chair was made, the age of the chair, etc. Instead, you focused upon one of the main *similarities* among almost all chairs: they are used for sitting.

This higher-level abstraction of the chair likely does not create any sort of stress between us, because the similarity you chose to focus upon is familiar to us both. However, by not also taking into account the *differences* my chair features compared to other chairs, we are not fully examining the multiple "levels" of the chair.

Perhaps the chair could also be abstracted as an "antique," which could also then be abstracted as "an heirloom." We would still be talking about the same chair, however, we have shifted our focus from its *function* to its *form* which is another higher-level abstraction.

These different levels of abstracted meaning could be described as "lenses" through which we discuss and interpret the chair, and in turn, the different lenses necessarily affect the ways in which we think about and interact with the chair.

For example, I may use language that relates to the history of the chair's construction and maybe even label it as a specimen from a particular school or era of design. As a result, I will also likely sit on the chair differently than I might a newer Lazy-Boy chair due to its design and value. My particular abstraction of the chair as an antique leads to particular behaviors and responses.

The same phenomenon would be true if I described the chair as a "hand-me-down." The term "hand-me-down" may suggest lesser value or even charm, which, in turn, may prompt me to plop down more carelessly on the chair than I would something described as an "heirloom."

Which label is correct? Heirloom? Hand-me-down? It depends upon one's *perception* of the object, as the words are not fully the things they represent.

When we ignore differences and only view something

through one lens, our behavioral, emotional, and spiritual responses are equally limited. Those limitations can also lead to skewed perceptions and inaccurate conversations that cause stress, fear, and other negative reactions.

On the other hand, by attending to the differences in information, sensations, outcomes, etc. of a situation, we can talk in ways that allow us to view situations through a variety of lenses. Some lenses may be negative, some may be positive, and some simply may not matter. But by purposefully looking for more than what initially meets the eye, opportunities for discovery through discussion come to life.

We can also look at life through a variety of lenses at the same time. For example, I could view my chair as an heirloom, but that then means it also has to be some form of a hand-me-down (that's how heirlooms become heirlooms). At the same time, I could also view it as my grandmother's favorite chair, and certainly it is still something to sit upon. All of these lenses (abstractions) for my single chair shape the way I think about it and use it on a daily basis.

The need for abstraction can be especially critical in illness settings. The overwhelming nature of medicine unfortunately causes patients and their circles of support to rely on high-level abstractions as they try to understand their illness.

Unless one has a strong medical background, trying to collect and discern information can be an enormous task. Many turn to the Internet to find answers, and in doing so, they are often unwittingly introduced to the need for abstraction.

Initial Internet search efforts often feature patients entering the name of the illness or disease they heard from the doctor. The search engine then "categorizes" available information into "hits" that are *similar* in terms of their use of the name of the illness or disease.

Unfortunately, because patients entered such a high-level abstraction term, they are inundated with information from hundreds of different sites that may or may not be helpful, let alone actually relate to their particular experience.

To help refine their information search, patients may then use strategies to *focus upon differences* that relate to specific aspects of themselves and their diagnosis. For example, a patient diagnosed with Type 2 diabetes may use the Boolean search method by entering "Type 2 diabetes + life expectancy + women + African-American" into the search box. This more refined search will result in hits that view the illness from these various levels of abstraction (gender, race, a particular type of diabetes, etc.). Certainly hits from the first search may still appear in this second search, however, it is also very likely new information will be introduced by purposely selecting different levels of abstraction.

So is the second search more accurate than the first? Possibly. But the more important thing to take away from this example is the fact that different types of information were discovered by examining the illness through multiple, and lower, levels of abstraction (lenses). Being able to view one's illness from a variety of levels of abstraction creates greater opportunities to respond to the illness in ways that can be less stressful and create room for hope.

Below is an example of an "abstraction ladder" one could create for a life-limiting diagnosis of Idiopathic Pulmonary Fibrosis (IPF). Note how the higher-order lenses feature more similarities than differences, and as a result, move further away from the complex biological nature of the illness itself. Also, each abstraction level necessarily prompts distinct mental, emotional, and behavioral responses. As you read through the levels of abstraction, try to examine your own emotional reactions to the semantically different lenses.

Illness Abstraction Ladder
(read from the bottom, lower-level abstractions on next page up through higher-level abstractions on this page)

Abstraction	Interpretation
Purpose	Also a very high level abstraction that focuses upon the illness having led the patient to a "higher calling." Often includes advocacy, volunteering, and networking with organizations and their participants who seek to find a cure for the illness and/or provide support to those experiencing the illness. Abstraction focuses on similarities with organizational missions and participant goals.
Opportunity	A very high level abstraction that focuses upon positive outcomes she might experience in light of the negative life event. Could include growing closer to family, taking on projects or trips she might have otherwise put off, finding a renewed sense of spirituality, etc.
Wake-up Call	Abstraction shares similarities with other life events that forced her to re-examine past behaviors, beliefs, values, etc.
Disabled/Dependent	Blanket abstractions that focus upon similarities of the patient to other humans who have limited physical, mental, and/or emotional abilities. Disregards varying degrees of disability/dependency among this population, as well as possible accommodations that actually allow elements of one's lifestyle prior to diagnosis to continue.

Smoker's disease	Abstraction focuses solely on smokers diagnosed with the illness. Omits other potential causes, including certain medications, environmental and occupational pollutants, and experience with Scleroderma, Rheumatoid Arthritis, Lupus and Sarcoidosis diseases.
Fatal	Abstraction focuses on similar thoughts among majority of medical community in terms of "survivorship." Omits attention to discrete factors that impact upon time of death after diagnosis (i.e., age, stage of illness upon diagnosis, etc.).
Interstitial Lung Diseases	A general term that includes many types of lung disorders, including IPF (below). Abstraction focuses on similarities of symptoms/diagnosis of IPF to other lung-related diseases.
Idiopathic Pulmonary Fibrosis (IPF)	Diagnosis based upon a particular set of symptoms/tests assigned and interpreted (IPF) by a medical professional. Abstraction focuses on similarities of the illness's symptoms to other diseases bearing this name.
(No words suffice at this level)	The biological construct of the particular illness that is unique to the patient. Does not require one's perception or understanding for the illness to exist.

It is important to remember no particular lens or abstraction is correct or right. Regardless of how sensible or even comfortable an abstraction may seem, all abstractions are imperfect and limited because of their attempt to categorize, and therefore, must always be challenged. By solely focusing upon similarities, abstractions become blanket assumptions that simply cannot hold true. As importantly, abstractions can easily turn into stereotypes that bias and skew behaviors.

The "disabled" abstraction above offers a good example. What qualifies as being "disabled?" An inability to wash one's face? An inability to walk up and down a flight of stairs independently? What about one's ability to remember? When does forgetfulness become a disability? Is disability a long-term experience? Can one be disabled for less than an hour?

Looking at these questions, it could be suggested disability lies in the eye of the beholder. Yet, the medical community and other societal institutions have come up with definitions for disability that relate to delivery of services and benefits. Those definitions are designed to make things easier and more convenient for the regulating bodies. However, do those definitions really make sense when one recognizes they are limited to similarities?

I would say no. In fact, I would suggest definitions of disability that rely solely upon similarities are as inaccurate and limited as any other form of *stereotype*. They do not take into account the individual differences that matter just as much as the similarities.

Furthermore, think of the emotional and physical reaction you have to the word "disabled." It would be safe to say if I announced to my best friend "I am disabled," both of us would communicate and behave in different ways. My friend may experience pity for me, seek ways to help me with daily needs, etc. I might experience feelings of helplessness, frustration, gratitude, etc. My use of that higher-level lens necessarily changes our behaviors and feelings, rightly or wrongly so.

Now imagine my friend's reaction if I were to instead say something like "If you have time, I need your help with my grocery shopping this afternoon." Would that statement suggest I was disabled? Possibly. But this second statement is simply more accurate in that it better explains what assistance I need without labeling me in some way. My avoidance of the "disabled" abstraction creates a whole different dynamic between us.

One of the more difficult abstractions in illness settings relates to patients abstracting their identities in response to the ways illness affect them. Each of us adopts a variety of personal and professional roles that dictate expected behaviors and ways of communicating (i.e., parent, boss, Little League coach, board member, sister, daughter, etc.) And we invest a great deal of importance into our various roles—so much so that they literally influence our behaviors and responses to others.

There are a number of challenges related to our assuming roles in life, regardless of the presence of illness.

First, roles are especially susceptible to grand assumptions and judgment as to whether or not they are performed well.

Second, roles can easily turn into stereotypes that are supported by one's culture and greater society.

And more dangerous still, roles are rarely if ever described with attention to the possibility of illness or disease. Roles are assumed to be what healthy and fully-functioning people "do."

Think about the role of a "parent." This term is an extremely high-level abstraction that is not only deeply rooted in our personal family experiences, but also in societal and cultural expectations. When we hear this category, each of us attaches our own unique baggage of similarities.

For example, we assume the person is either a mother or father, is responsible for teaching children right from wrong, is responsible for feeding and clothing their children, etc. These are the *similarities* we tend to observe among people described as parents.

Few of us would likely suggest a parent is one who is sick or one who is terminally ill. That wouldn't make sense, because then how would they be able to physically pick up their children or help chaperone the 4th-grade field trip?

Yet there *are* parents who are very sick and unable to do those activities and more. Does that mean then the parent who has cancer is not really a parent? Certainly not, but because we all buy into high-level abstractions when it comes to our roles, it can seem like the answer to the question is "yes." We forget while one's body enables one to perform the expected behaviors and thoughts of various roles, many of those roles can still be maintained or accommodated in the wake of serious illness.

When patients review their roles in life, it is critical for them to remember the limitations and inaccuracies of those roles in the first place. In many instances, patients will need to re-imagine how those roles need to be modified, eliminated, or replaced.

Their circles of support will also need to do the same. Everyone affected by an illness experience needs to re-examine how roles can and do come to life in both sickness and health.

On the following pages is another possible abstraction ladder that examines diverse roles for a patient diagnosed with Stage IV breast cancer. Note how the similarities of the higher-level abstractions move us further and further away from the actual person to whom we are referring.

Abstraction Ladder for Roles of a Woman Diagnosed with Breast Cancer

(read from the bottom, lower-level abstractions on next page up through higher-level abstractions on this page)

Abstraction	Interpretation
Citizen of United States	Abstraction shares similarities with other citizens. Omits key characteristics unique to the patient.
Christian	An abstraction that focuses upon similarities to a religious mindset shared by others who also label themselves as Christians. Omits characteristics unique to the patient's faith development and beliefs.
Breast Cancer Patient	Blanket abstraction that lumps the patient with all other people diagnosed with breast cancer. Omits the fact her experience with breast cancer cannot be identical to any other person's experience with breast cancer.
Wife	Abstraction focuses on similarities with other "wives." Also highly ambiguous as noted below.
Working Mother	Abstraction focuses on similarities with other "working mothers." Also highly ambiguous as noted below.
Mother	Abstraction focuses on similarities with other "mothers." Highly ambiguous, as the term is not only influenced by personal experience, but also societal and cultural norms.
"Sandra"	Name given to this particular female patient. Suggests individuality, however, the word is not the thing, so the name *stands for* the patient. As a result, any and all distinguishing factors of this particular female patient are left out when this reference label is used. The name is applied in a holistic fashion as a label.

Female human	Abstraction that focuses upon similarities based upon gender within the human race. Does not allow for unique characteristics particular to the patient.
Human being	Abstraction that focuses on similarities with other human beings largely at biological and mental levels. Omits unique characteristics particular to the patient, including biological differences related to health status.
(No words suffice at this level)	The biological construct at the molecular level that is unique to the patient. Does not require one's perception or understanding for the patient to exist at this level.

We all can find comfort in the roles we choose for ourselves, just as we find comfort in rejecting the roles we do not want. But those roles upon which we rely are loose at best. We must remember they are lenses that focus and limit the ways in which we view ourselves, and while they come about largely through our own personal experiences, they are also shaped by our society and culture over time. Roles constantly change and evolve, albeit some more rapidly than others.

Illness certainly prompts examination of those roles as patients seek ways to adjust and accommodate to new lifestyles. By examining the words we use to describe our roles in life (our lenses), we can hopefully pursue conversations that relieve the ambiguities of those roles and make room for multiple interpretations.

The following questions can help us work with abstractions in our conversations with ourselves and others:

- Am I clinging to higher-order abstractions that never fully fit in the first place?

- Where do my abstractions come from? How have my personal experiences informed my abstractions? How has society and culture informed my abstractions?
- What "differences" about myself, others, situations, etc. am I ignoring or forgetting in my abstractions? What happens if I focus upon those differences?
- How can I describe myself outside of specific roles or other abstractions?
- How can I modify the way in which I talk about the roles I choose to adopt?

"To be" or not "to be."

Before diving into this next strategy, take a moment to complete the following exercise. You will need paper and a writing instrument.

> First, think of someone you absolutely cannot stand—someone you hope to avoid if at all possible. Now write a brief description of the person you hope to avoid.

> Second, think of someone you absolutely enjoy spending time with—someone you look forward to seeing. Again, write a brief description of the person you so enjoy.

> Finally, circle all the "is" or other "to be" words (was, were, are, will be, etc.) you included in each description.

If your descriptions are anything like mine, you likely used some form of the word "is" regularly throughout both descriptions. For example, my first descriptor about a coworker features sentences like "He *is* a jerk," "He *is* obnoxious," "He *is* the laziest person on earth." My second descriptor about my girlfriend includes an equal number of "is" references: "She *is* the kindest person I know," "She *is* Superwoman," "She *is* funny."

Now notice how my use of the word "is" in my statements creates an almost mathematical equation in which "is" serves as an equal sign.

He = jerk
He = obnoxious
She = Superwoman
She = funny

This *is* where the danger lays, because our word equations can easily lead to bias and closure in our minds, as well as the minds of those with whom we are talking. This is especially true for conversations in which only one of the speakers is familiar with the topic being discussed. The receiver of that information usually has no way to qualify the information when it is shared (at least initially).

I'm guessing you don't know my coworker personally, so all you can go on is what I tell you. Since I told you he was a jerk, it is likely you would expect to meet some boorish fellow.

But what if after actually meeting him you found him to be a nice and interesting person? Awkward, right?

Similarly, what if after meeting my girlfriend, you actually found *her* to be obnoxious and not at all funny? You are now faced with the task of trying to wrestle with the faulty equations I planted in your mind. Was I not telling the truth? Are you missing something? Etc.

Now here is some additional information about my friend: she *is* an accountant. What do you think of her now?

Perhaps you think of someone who dresses conservatively, someone who is good at math, and someone who tends to be soft-spoken. *Is* my friend that way? Possibly, but even if she does demonstrate those behaviors, they may not relate to her being an accountant, and there is also certainly *much more* to my friend than her job.

What if my friend does not wear suits or she is not quiet? What if she also races stock cars? Does that make her less of an accountant? Certainly not, but our semantic connections to the label "accountant" did not allow us to easily consider other possibilities.

Furthermore, because I *only* spoke of my friend in terms of her career, I forced you to focus on that single aspect, regardless of the role her career may actually play in her life. I did not give you anything else to go on, so you were left with an incredibly incomplete picture of my friend. Still, we both assume we *know* my friend.

Trying to have a conversation without using the word "is" poses a real challenge, especially when we are talking about ourselves or others. But there *is* hope! We can add "qualifiers" to what we say that leave room for other possibilities.

Qualifiers include phrases like "seems to be," "appears to be," "seems like," etc. By including qualifiers in our statements, we suggest we are reporting from our limited *perception* versus reality.

Consider the set of sentences on the next page. The first set presents an "is" statement about Michelle. Note how that report comes across as both judgmental and final. It makes Michelle appear one-dimensional in relation to a clinical diagnosis.

On the other hand, my reframed statements seem to leave room for the possibility of *other* reports about Michelle. They are more informative than judgmental, and as a result, they create opportunity for further discussion.

Their use of qualifiers also leaves room for the possibility of my being mistaken, which again leads to opportunity for further discussion and clarification.

Original "is" statement

Michelle *is* a breast cancer patient.

Reframed statements

Michelle received a diagnosis of breast cancer from her doctor last May.
It appears Michelle's chemo treatments will prevent her from continuing with our book club this month.
Michelle seems to be volunteering more with the Susan G. Komen Foundation event this year.

Each of the above reframed statements *could* suggest Michelle *is* a breast cancer patient, but note how they more so emphasize what I have observed, and as a result, help prevent us from making judgments about Michelle as a person. None of my statements blanket Michelle with some particular label or role. They instead report about specific actions, behaviors, and information I either observed about or heard from Michelle.

The statements also leave room for more to be said about Michelle, and removing the label of "breast cancer patient" also removes any sense of closure or finality about who Michelle *is*. My reframed statements position cancer as representing a part of Michelle's life versus her whole being.

And the third sentence could actually describe *anyone*, as one does not have to have breast cancer to support the foundation.

Use of the qualifiers "appears" and "seems" also leaves room for my observations to be modified in case new information or new observations are introduced at a later point in time. After all, how would I really know if my observation about chemo treatments being the cause of her not attending book club meetings is accurate? Might she have simply had a change in her schedule for some other reason?

In terms of her volunteering, unless I can compare her actual number of volunteer hours then and now, that statement is also ripe for error. Maybe I sense she is volunteering more because I wish I could spend more time with her, as our friendship has grown since her diagnosis. In fact, she may be volunteering the same amount of time, but I just notice it more this year.

By using the word "seems," neither I nor Michelle are locked into faulty assumptions, and we can imagine other possibilities and options, if needed.

Following are additional examples of how to reframe an "is" word equation. As you read each set, note your physical and emotional reactions to each statement in terms of whether or not you sense closure or judgment in the report. Also notice how the reframed statements create room for other interpretations, and as a consequence, other possible responses and behaviors in response to what is said.

Original "is" statement

Cancer *is* a death sentence.

Reframed statements

Some people die as a result of complications from cancer.
Some forms of cancer seem to be more life-threatening than others.
It seems many people diagnosed with my same form of cancer experience remission at some point in time.

Original "is" statement

The doctor *is* an expert.

Reframed statements

The doctor has received several recognitions from important medical organizations for her work.
The doctor has successfully performed over 100 similar operations.
It seems both patients and professionals often turn to this doctor as a resource regarding this particular illness.

Original "is" statement

I *am* a victim of Lou Gehrig's disease.

Reframed statements

I have been diagnosed as having ALS.
Complications from my ALS make me feel scared and helpless at times.
It seems ALS occurs throughout the world and has no racial, ethnic, or socio-economic boundaries.

Think of ways you might reframe the following "is" equations by eliminating the word "is" and adding qualifiers for clarification. Try to avoid judgment or closure and leave room for other possibilities.

- The wheelchair *is* an embarrassment.
- The treatment *is* the solution.
- My family *is* my support system.

In addition to being mindful of "is" equations that relate to one's identity, we also need to be mindful of using "is" equations that feature adjectives on the right-hand side of the equation (remember, adjectives describe people, places, and things).

Adjectives tend to introduce an element of judgment (good or bad, right or wrong, worthy or not worthy, etc.) that complicates communication.

Adjectives also tend to be hugely ambiguous, and the personal spins we attach to adjectives ranges widely. Shared meaning is especially difficult to achieve in this regard.

And finally, adjectives do not allow for change. Instead, they suggest something *is* and always *will be* so.

> **Examples:** Dan (person) *is* sick (adjective).
> My mother (person) *is* suffering (adjective).
> The nurses (people) *are* supportive (adjective).

In terms of the first report about Dan, what does "sick" mean? For some people, a diagnosis is enough to make someone sick. If a disease or illness has been identified, then that person must *be* sick, regardless of any observed presence of symptoms. Others view sickness as the times in which one experiences physical pain or discomfort, like having the cold or flu. Still others view sickness in terms of how much it disrupts one's life. One can be described as "sick" or "really sick" based upon how much or little one's routine is impacted by an illness.

The word sick *is* ambiguous and the concept ultimately lies in the eyes of the beholder. Herein lays the problem.

When we refer to *Dan's* being sick, what exactly are we talking about? Most of us would at least consider his physical signs of sickness (i.e., fever, vomiting, pain, etc.), but *is* Dan simply the sum of those experiences? Might there be times when Dan is not feverish or nauseous? Certainly, so what would we call Dan during those times?

And what about the rest of Dan? Are his thoughts sick? Are his emotions sick? Is his spirit sick? Likely not, but our using the single descriptor of "sick" prevents us from thinking about these questions about who Dan *is*.

The word "sick" also introduces a level of judgment about Dan in that we view sickness as a bad thing. Who of us wants to be sick? Unfortunately, our negative reaction to the word can easily be transferred to Dan the person if we only use the word "sick" to say what he *is*. Our behaviors toward Dan will then also be influenced by that extremely limited description. We may avoid visiting Dan, we may fear Dan, we may pity Dan, etc. By describing Dan solely as being "sick," we will treat him differently than we might if we remembered the word is not the thing.

The same negative consequence could happen from Dan's viewpoint. If he defines himself solely as "sick," his behaviors and assumptions will be shaped (and limited) by his take on the word. Perhaps he will avoid friends, stop exercising, cut back at work, etc. Like us, Dan will forget there is more to him than "being sick," and therefore, he will limit his thoughts and reactions in direct proportion to the limits of the word "sick."

The word "suffering" is also hugely ambiguous and packed with semantic baggage. Our thresholds for physical and emotional pain and discomfort are unique to all of us. What might constitute "suffering" for me will certainly be different from any other person's interpretation of "suffering." What *is* suffering? When does it begin? When does it end? How does the patient's suffering relate to caregivers' suffering? These questions point to the inadequacy of the word to fully describe the lived experience and its inability to address changes over time.

The inadequacies of the word "suffering" may also prompt inappropriate responses and behaviors. Think of how we as humans respond to those who are suffering. We are often moved to thoughts of pity and sadness that influence our response to that person. For example, we may offer extra help, we may express our condolences, we may become overwhelmed at what we cannot fix, etc. — all based upon our solely describing a person as suffering.

What if that person does not view herself as suffering, yet we still demonstrate those behaviors? It is likely she would become impatient, and possibly insulted, by our behaviors. We, in turn, would be surprised at her reaction to our wanting to help. Ultimately, we could both feel frustrated, angry, resentful, etc.

The same problem could arise if the patient did view herself as suffering, however, her caregivers did not. Both the patient and her caregivers could easily grow frustrated and resentful. The patient might think the caregivers did not believe her pain was as incredible as she described, and were therefore, unwilling to help reduce it in any way. On the other hand, the caregivers might view the patient as not owning up to the inevitable pain a diagnosis bears (especially if the patient refuses pain treatment).

To stem these dilemmas we need to avoid using single adjectives or short adjective phrases to describe people, places, things, ideas, etc. We should instead try to only describe observed behaviors and share information that is time-stamped through the use of qualifiers (i.e., "it seems," "it appears," etc.).

Original "is" statement (using an adjective)

I *am* independent.

Reframed statements

I try to do as much for myself as I can.
I value independence.

There are times I need to rely upon others, but for the most part, I have been able to cook and clean on my own.

Original "is" statement (using an adjective)

My wife *is* frustrated.

Reframed statements

My wife seems to sorely miss being able to drive.
My wife appears to have great difficulty getting dressed.
At times, I see my wife crying, and I don't know why.

Original "is" statement (using an adjective)

My prognosis *is* good.

Reframed statements

Test results show the tumor appears to have decreased in size.
The doctor suggested early detection allows us to have several options for treatment.
Many people who receive the same test results have a good recovery rate.

Original "is" statement (using an adjective)

My life *is* different.

Reframed statements

I have made accommodations to my changing abilities.
I think about things I never did before.

I work from my home more often than I did in the past.

If you find it hard to let go of adjectives, try to at least replace the "is" words with verbs that can help reduce hidden assumptions, biases, and stereotypes.

Using the examples above, instead of saying "Dan *is* sick," you could instead say "Dan *looks* sick" or "Dan *appears* sick." For the nurses' example, instead of saying "the nurses *are* supportive," you could instead say "the nurses *act* supportively" or "the nurses *offer* support."

The reframed statements not only leave room for Dan and the nurses to be more than that single adjective, but they also leave room for the possibility your information is inaccurate or incomplete. Neither Dan nor the nurses are locked into the single adjectives as a whole.

In summary, trying to eliminate use of "to be" words in any shape or form is difficult at best. It is likely we will not be able to completely eliminate use of those verbs, however, it behooves us to at least start evaluating our use of those verbs when we are feeling stress, anxiety, frustration, etc in our conversations.

"Is" equations are essentially untruths due to their limited scope and lack of attention to the fact things change over time. Those untruths can be especially dangerous if we invest in them to any real extent, because they force our thoughts and behaviors to be inaccurate and inappropriate to the situation.

We need to develop ways of talking about ourselves and our lives in ways that leave room for possibilities versus faulty assumptions of certainty. To help us better communicate in ways that recognize life as a process over time, we should ask ourselves the following questions:

- Can I really say what something *is* in its totality?

- How am I limiting my observations by using the word "is" or other "to be" verbs? What am I leaving out? What more could be said?
- Am I reporting in ways that leave room for change or misunderstanding on my part? Do I leave room for the introduction of new information over time?
- What judgments or biases do my word choices introduce? How can I modify what I say to reduce or eliminate those judgments?

Thoughts on "hope" and "faith"

I have gained confidence over time in using these "word is not the thing" strategies, and as a result, I have had rich and meaningful conversations with family and friends.

I have also found these strategies to be especially helpful when working with patients and their circles of support. I have been honored to be a part of many conversations in which the strategies led to creation of meanings that fully empowered people and also helped them find peace.

Still, I have not been able to easily apply the strategies to two key words: hope and faith.

The strategies shared in this book so carefully stress our speaking about *that which can be observed* versus relying upon assumptions or judgments. We are supposed to talk about that which we directly experience so we can better say it like it is.

So then how do hope and faith fit into the picture since neither is necessarily *observable*, let alone prone to some form of "proof?" Faith especially is based upon things unseen or unknown — that's what makes it "faith."

For many patients and caregivers, hope and faith are critical to maintaining sanity in the wake of seemingly uncontrollable events in their lives. Patients and caregivers cling to hope and faith's ability to simply help them get up and face each day. Perhaps today will be better. Perhaps a cure will be found. Perhaps there is something to learn from all of this.

While I have come to the conclusion that "word is not the thing" strategies can relate to the words hope and faith, I would also seek exceptions to those strategies that allow people to use the words as if they *are* the things they represent. Recognizing the immense power these two small words hold in illness settings, I hesitate to seek alternative ways of discussing hope and faith.

In terms of words not being able to say it all, it is true each small word is hardly up to the task of representing the infinite number of thoughts and dreams that can change over time. One can never fully define or explain what one's hope or faith *is*—where it comes from, what it addresses, when it is in place, when it is wavering or lost, etc. Hope and faith are extremely personal experiences which defy explanation.

Still, hope and faith are incredibly important concepts to imagine and discuss in illness settings for both patients and caregivers. There is something physically and emotionally cathartic that comes from using those two exact words in conversations, regardless of their inability to fully represent the experiences to which they are assigned. The words themselves have incredible influence on thought and behavior.

> "I have *hope* this treatment will work."
> "I have *hope* a cure will be found soon."
> "I have *hope* my family will support me."
> "I have *hope* I will be able to adjust."
>
> "I have *faith* there is something to learn from all this."
> "I have *faith* death is not the end."

> "I have *faith* that God will help me through the tough times."

Because of each word's semantic backdrop, I suggest we be allowed to use the words in our conversations without having to define, defend, or otherwise explain what those terms mean to ourselves or to others. Yes, the words will never fully be representative, but that will be okay in light of the important semantic role they play in illness settings.

Instead of concerning ourselves with finding replacement words or phrases that may be more descriptive or accurate, I think we should allow ourselves to use the words hope and faith with the understanding they relate to any and all facets of an illness experience.

We can also attend to the contextual dimensions of using the words hope and faith, especially in terms of one's religious or spiritual beliefs. If we know the speaker believes in a spiritual resurrection upon her death, we can better understand her faith that she will be rejoined with family members who predeceased her. Similarly, if we know the speaker believes in the power of prayer, we can better understand her hope that her prayers will be answered.

For these religious or spiritual backdrops of hope and faith, I think it is better for us not to challenge each other's positions or ask for one to defend her beliefs. Instead, our conversations should make room for hope and faith to be discussed openly and without judgment while remembering the words are never fully the things they represent. I cannot fully understand your levels of hope and faith any more than you could fully understand mine, but that does not negate their existence.

Contextual dimensions also point to our need to remember levels of hope and faith can change over time. Neither is easy to maintain when one experiences personal setbacks or physical decline, so one's levels of hope and faith can fluctuate.

One way to address that fluctuation is to remember to replace "to be" words. Instead of saying "I *am* hopeful" one could instead say "I *have* hope today."

Likewise, instead of saying "I *am* faithful" one could instead say "I *have* faith today." Replacing "am" with "have" leaves room for the times in which one's hope and faith can wane, and time-stamping the statement ("today") also leaves room for times of doubt, frustration, etc. By saying one "has" hope or faith, room is left for *other* feelings or beliefs to also be experienced.

In terms of abstraction, hope and faith are extremely high-order lenses. They can include any and all responses to any and all events that the speaker *interprets* as being hopeful and/or requiring faith. Recall that strategies shared earlier in this chapter would have us avoid reporting such limitless categories as if they were actual singular experiences, however, it is that limitless nature to which patients and caregivers cling. Trying to describe those limitless concepts as specific events or behaviors is not only frustrating, but also seemingly impossible. Hope and faith just *are,* and ironically, their higher-order level of abstraction seems to bring patients and caregivers more comfort than does any concrete and limited description of personal experience.

Therefore, I would suggest another exception be made that allows us to use the words hope and faith in our conversations as the higher-order abstractions they represent.

I also wrestle with the notion of "roles" related to being hopeful or faithful. Is "being hopeful" just as much a role as being a "parent" or "teacher?" If so, the same types of dilemmas are imposed.

For example, what *is* a hopeful person? What do they *do* exactly? Do they smile in the wake of pain? Do they say things like "all will be well in the end" or "I'm confident a cure will be found"? What if a person does not smile in the wake of pain or say those things? Is she not a hopeful person? What if she has crying bouts or lashes out at a good friend? Has she lost hope?

It is also especially important for caregivers, friends, and family members to avoid assigning the roles of "being hopeful" or "being faithful" to each other. I often hear friends and family describe people who are ill as *being* hopeful or faithful: "she is so hopeful" or "he is so faithful."

Assigning such roles to patients (or caregivers) does not provide room for patients to *not* be that way, and frankly, imposes a heavy burden upon patients and caregivers to help those of us who are *not* ill to feel better about their situations. We want patients and caregivers to have hope and faith because it helps *us* have hope and faith. It can be upsetting to witness a patient's sudden bouts of depression or a caregiver's short temper when we thought of them as hopeful and faithful.

But the fact is, hope and faith fluctuate over time. Not ascribing hope and faith as roles gives people permission to experience the ups and downs of their illness journeys without feeling guilty, stressed, frustrated, etc

In closing, it is very possible my concerns about the impact of our using the words hope and faith are unfounded, but because of the pivotal role those two small words seem to play in illness settings, I thought it important to explore. I know the power the two words have in my own life, and I would feel bereft at having to replace or eliminate their use. The words in and of themselves give me comfort in challenging times, and in that regard, I choose to make exceptions, as needed.

Chapter 3

The Map is Not the Territory

"The worlds we manage to get inside our heads are
mostly worlds of words."
-- Wendell Johnson[1]

"The only man who behaves sensibly is my tailor; he takes
my measure anew each time he sees me, whilst all the rest
go on with their old measurements and
expect them to fit me."
-- George Bernard Shaw

> "The doctor said Sharon stands a small possibility of beating her illness if she pursues radiation. But if she doesn't pursue the treatment, the doctor is pretty certain Sharon will only get worse, and we'll lose her."
>
> "Didn't that treatment work last time? I thought she went into remission."
>
> "Yes. But for some reason, the cancer has come back."
>
> "I'm so sorry to hear that. Sounds like neither choice is for sure."
>
> "Yeah. We always seem to get bad news."

Before we examine the above conversation, let's first talk about "maps."

Maps can and do help us navigate. Anyone who has planned a summer road trip or backcountry hiking event understands the importance of accurate maps. And the explosion of reliance on global positioning systems reflects humans' desire to know where to go and how to get there fast and efficiently.

Maps give us comfort. They help us plan and predict. They can even warn us of problems ahead.

But those of us who have lived for more than a few decades know maps *change over time*. Countries are reshuffled and reconnected. Old highways are abandoned and new superhighways are built. Businesses along Main Street come and go.

The same phenomenon occurs with the *verbal* maps we create for ourselves. Just as our geographic maps become outdated, so, too, do our verbal maps.

Read through the following verbal maps and think about your knee-jerk response to each:

- You need a college education to get a good job.
- You shouldn't swim right after you eat.
- Welfare is a sign of dependency.
- Women share their feelings more easily than men.

- You can't have your cake and eat it too.

I'd bet many, if not most, readers would initially agree with the above statements because they represent maps that have become ingrained in our society as a whole and for many of us as individuals.

However, isn't it true exceptions can be found for each of the above maps? Of course. But our first impulse is to agree because it is *easier* to do so. The verbal maps are what we *know*, and the longer we have used those maps—and the longer we have not been introduced to contradictions to those maps—the more easily we rely upon those maps.

Now examine the following verbal maps related to illness settings:

- Medical specialists are smarter than regular doctors.
- Starve a cold, feed a fever.
- Pain is just something you live with.
- Family members are the best caregivers.
- Test results don't lie.

Again, I'd bet most readers would find truth in several of the above maps because they, too, have become steadfast in illness settings. And while it is also quite possible to find exceptions to all of the above maps, most of us are hard-wired to accept more easily than to deny the statements.

It *is* hard to identify and develop new maps, especially the longer we live. Unchartered territory can be frightening to say the least, so we often desperately cling to familiar maps hoping they will somehow fit new situations. But that rarely happens, and as a result, we can suffer from disappointment, anger, and other negative feelings.

It is also hard to know what we don't know. Because so much of what we believe appears "factual," purposely questioning what we think we already know takes serious creative energy. Why would we question ourselves? We know what we know…period.

Here's a challenge: come up with a new melody for the "Happy Birthday" song.

For me, that task would be incredibly daunting and uncomfortable at best. First, it is very difficult for me to even imagine something other than that melody. That *is* the melody, because that is how it has been sung since the beginning of time (or at least my time).

Secondly, most people would think I had lost my mind trying to change the *perfect* birthday song. Everyone sings "Happy Birthday" that way!

Third, I *like* the original melody. Why would I want to change it?

While this is a simplistic example, it demonstrates the power a verbal (and musical) map can have over our ability to think in new ways.

"The map is not the territory" principle tells us it's okay to rely on verbal maps as long as they reflect reality, however, knowing change is constant, we must also remember even the best-crafted maps cannot stand the test of time. Therefore, we need to create maps that are as accurate as possible in the first place and then constantly revisit those maps to see how they are affected by changing times.

Getting past "either/or" perspectives

This chapter's opening conversation features a familiar verbal map of life with cancer that features two destinations: "the land of treatment" or "the land of no treatment." As the friend points out in the discussion, neither destination has any certainty to it, so the dialogue's overall tone becomes less than hopeful.

To make matters worse, the speakers' either/or approach to Sharon's illness is not only stressful, but it is also inaccurate.

Certainly treatment is a pivotal part of Sharon's cancer experience, but it is not the *only* part of the cancer experience—nor is it the *whole* territory. Other activities might also influence Sharon's illness experience, including diet and exercise, laughing with family and friends, and meditation and prayer.

The possibility also certainly exists that Sharon could live beyond doctors' or anyone else's expectations *without* pursuing treatment, yet that destination is not naturally featured on the map. As a result, hopes are limited to the two destinations and worry sets in.

Healthy conversations in illness settings require us to move from our either/or perspectives to a more multi-valued approach to life that helps us avoid getting stuck or lost. Talking about an illness experience as literally filled with possibilities, some of which are not as dire or threatening as others, can help reduce stress and anxiety and create room for more thoughtful responses.

Below are examples of common either/or orientations in illness settings. As you look at the pairs of options, ask yourself are they really mutually exclusive and/or are they really the only or best options.

"Either/or" orientations in illness settings

natural remedies/pharmaceutical remedies
independence/dependence
good news/bad news (related to a diagnosis)
pain/no pain
pursuit of treatment/hospice care

In terms of the "independence/dependence" dichotomy, many patients fear losing their independence as they manage their illness, so they resist asking others for help along the way. But most patients' need for assistance actually fluctuates across their illness journeys.

For example, a patient may need rides to and from the clinic during her chemo treatments, but she may not need rides for her follow-up appointments after treatment. Or she may be able to manage her own grocery shopping before a chemo treatment, but she may need someone to run to the store to pick up a prescription after the treatment. I'm betting few of us would label such instances of receiving help from others as suggesting the patient is fully dependent upon others, especially recognizing how much the patient does herself outside of those activities. The either/or scenario is simply inaccurate.

The limitations of an either/or approach are also apparent in terms of the "good news/bad news" dichotomy, especially related to test results. Is a higher test score better? Is a lower score worse?

We must first remember test results reflect whatever was going on in our body during the hour and day in which we took the test. Certainly a test result can be highly accurate, but it still only reflects a moment in time. Our bodies continue to change, so there is always the possibility a test result could become obsolete over time.

Test results are also limited to measuring very specific things. For example, a blood test can only measure characteristics found in the blood. It cannot measure other fluids in our body, electrical impulses in our body, etc. Physical tests also cannot measure our emotional or spiritual health. Therefore a test result can only tell part of our story.

What I'm suggesting is that test results are a snapshot in time that may or may not remain constant. And knowing the possibility for change always exists, describing a test result as "good news" could set us up for disappointment if later test results paint a less rosy picture.

That does not mean patients and caregivers shouldn't celebrate a test result that shows signs of healing and improvement. But they should try to avoid labeling test results as solely "good" or "bad." They need to instead make statements with qualifiers that recognize the possibility for change in the future.

Either/Or Statements

"My blood test for anemia was good." (good vs. bad)
"The biopsy report was all bad news." (good vs. bad)

Reframed Statements

"Yesterday's blood test suggested I showed no signs of anemia."
"The biopsy test I took last week showed the presence of some cancerous cells."

Trying to view test results through multiple lenses can be incredibly difficult, but trying to view an actual diagnosis as anything other than "bad news" may seem absolutely impossible. How could a diagnosis of cancer, heart disease, infection, or other malady possibly be viewed as anything other than tragic?

To suggest there could be benefits from a life-limiting diagnosis seems ludicrous to most of us, yet a growing body of research housed within the communication, psychology, and philosophy disciplines is doing just that. Researchers are suggesting it *is* possible for humans to find positive meaning in the wake of negative life events.

Dr. Viktor Frankl[2], an Austrian neurologist and psychologist, examined Holocaust survivors' abilities to face the nightmare of concentration camps in ways that actually helped them find hope and promise. Frankl had been a prisoner himself, along with his father, mother, brother and wife. Frankl was the only one to survive the camp.

Frankl later reflected upon the ways in which fellow camp members coped (or not) with the horrors they faced each day, and then he suggested that facing a tragedy that cannot be avoided actually provides humans the ability "to witness to the uniquely human potential at its best, which is to transform a personal tragedy into triumph, to turn one's predicament into a human achievement."

Dr. Peter Koestenbaum[3], a philosophy professor at San Jose State University in the 70s, also examined the potential benefits of viewing a life-limiting illness and/or imminent death through multiple vantage points. Koestenbaum developed a series of exercises that prompted patients to reflect upon their thoughts and feelings related to their past, their future, their sense of self, and their relationships with others. Participants reported such purposeful reflection across a variety of levels enabled them to find meaning in their illness experiences and ultimately achieve a sense of peace.

Several memoirs have also been written by those facing serious diagnoses that describe new-found meanings authors discovered within their illness experiences. For example, Jenifer Estess's book[4] *Tales from the Bed: A Memoir* chronicles her experience with ALS and how it drove her to create the "Project A.L.S." organization.

Reynolds Price's memoir[5] *A Whole New Life: An Illness and a Healing* describes his experience with a cancerous tumor in his spinal cord (which ultimately caused him to become paraplegic) that led him back to a prolific writing life.

And Dr. Oliver Sacks's book[6] *A Leg to Stand On* reports his healing experience from a terrible injury to his left leg while hiking in Norway. The experience forced Sacks to become the patient (he is a renowned neurosurgeon) who had to confront the daily calamities of medical care and physical disability. He suggests that experience helped him become a better practitioner as he gained new-found insights into the meaning of illness and disease.

One of my favorite illness memoirs is Eugene O'Kelly's *Chasing Daylight: How My Forthcoming Death Transformed My Life*[7]. The former CEO of KPMG wrote the powerful little book over the three and a half months between his diagnosis and ultimate death from brain cancer. The excerpt below from the opening chapter entitled "A Gift" showcases his extraordinary response to his diagnosis:

> "I was blessed. I was told I had three months to live.
>
> You think that to put those two sentences back to back, I must be joking. Or crazy. Perhaps that I lived a miserable, unfulfilled life, and the sooner it was done, the better.
>
> Hardly. I loved my life. Adored my family. Enjoyed my friends, the career I had, the big-hearted organizations I was part of, the golf I played. And I'm quite sane....
>
> I was forced to think seriously about my own death. Which meant I was forced to think more deeply about my life than I'd ever done. Unpleasant as it was, I forced myself to acknowledge that I was in the final stage of life, forced myself to decide how to spend my last 100 days (give or take a few weeks), forced myself to act on those decisions.
>
> In short, I asked myself to answer two questions: *Must the end of life be the worst part?* And, *Can it be made a constructive experience – even the best part of life?*
>
> No. Yes. That's how I would answer those questions respectively. I was able to approach the end while still mentally lucid (usually) and physically fit (sort of), with my loved ones near.
>
> As I said: a blessing."

It is difficult for me to attempt to offer strategies that can quickly or easily allow us to turn our negative paradigms about illness and disease on their heads. Who of us knows if we will be able to craft such a multi-valued outlook on an overwhelming diagnosis? Imagine being able to find something positive in a life-limiting diagnosis, let alone being *grateful* for such a report!

Yet many people have done just that, and I would say those people's ability to look at their illness experiences through multiple lenses enabled them to find authentic goodness and grace along the way.

One suggestion I offer to avoid either/or conversations is to talk about *living with* or *living in* an illness experience. Illness, disease, and injury are certainly not traveling companions any of us would choose on purpose. But once they appear in our lives, we will be forced to travel together for at least some period of time.

By making statements such as "I am living with cancer" or "I am living with diabetes," the act of living takes precedence over the illness or disease because it is said first, and therefore, considered first. Statements like these remind us we are living…we are still here…*because we said so.*

We are then welcome to find meaning in our newest journeys—personally, professionally, spiritually, etc.—just as we were able to find meaning in our lives before our diagnosis.

We all have the ability to find a variety of meanings within an illness journey if we are open to the areas that exist outside of "the land of treatment" or "the land of no treatment." We have multiple destinations to which we can travel that feature opportunities to learn more about ourselves and those who care about us if we choose to view our illness experiences through multiple lenses.

The following questions can help us purposefully seek out multi-valued orientations in an illness setting:

- What happens if I wait to react?
- What other options might I imagine?
- Am I limiting myself to two or less possible choices?

- Am I allowing myself to adopt different outlooks at different times?
- What else can I say about this situation?

An experience unlike any other

Illness journeys are rarely static; both patients and caregivers experience ups and downs along the way that are largely unpredictable. The challenge in that regard is to talk in ways that acknowledge and leave room for those changes to happen.

In the chapter's opening conversation, the friend wonders why the radiation that worked last time for Sharon would not be suitable for her newest cancer challenge. Unfortunately, it appears the landscape of Sharon's cancer experience has changed, even though it is technically the same form of cancer as her original diagnosis. That means the "remission" map related to Sharon's prior radiation treatment will no longer be accurate. Instead, she needs a new map and likely a new destination.

Simply put, $Cancer_1$ is not the same as $Cancer_2$. Even if the cancer is given the same name, the second experience will likely be very different in terms of Sharon's physical response to treatment, her emotional response to treatment, her husband's experiences, etc. So despite both journeys featuring cancer as the impetus, Sharon and her husband need to incorporate the use of an "indexing" strategy as they create new maps to guide themselves through this new experience.

Using words and phrases that help us "index" events and ideas is akin to abstraction noted in the prior chapter. We are challenged to *look for differences* among seemingly similar situations. Regardless of how similar an experience may seem to another, it is simply not possible for them to be identical. As the saying goes, time changes everything, so no two situations can be exactly the same.

Stress and anxiety are closely connected with the extent to which one's expectations are matched. When things work out as planned, we remain content. If things go even better than anticipated, terrific! But when things do not go as planned, we feel a variety of negative emotions. And the greater our expectations are foiled, the stronger the negative reaction.

For example, if Sharon expects to find a treatment that does not result in her losing her hair (like her first treatment experience), she will be disappointed to learn that option is no longer available, and her anxiety will increase.

On the other hand, if Sharon expects to experience uncomfortable sensations in her hands and fingers (similar to her first treatment experience), when none result, she will be relieved.

Obviously it is extremely difficult to be "pleasantly surprised" about any cancer treatment experience. But the point is that recognizing and talking about the differences across treatment experiences can help Sharon and her loved ones better cope with what actually happens.

Entering the experience without preconceived notions can also leave room for hope and more positive outcomes. Holding onto expectations that simply cannot happen causes stress, so Sharon and her family need to remind themselves along the way that when things do not go as planned, that is because it is a *different* experience. There is not necessarily a right or wrong to the matter. It simply is different.

The same can be said for hereditary illnesses. I had a friend who was diagnosed with Lou Gehrig's disease (ALS) when she was 60. She had already lost her son to ALS when he was 27, and her younger sister was showing signs of the disease as well.

Throughout our friendship, my friend kept comparing her experience with ALS to her son's, especially in terms of expectations of how quickly she would lose control of her abilities to speak and swallow. She literally clocked the changes over time and then grew agitated when her experience didn't mesh with her son's timeline.

I think all along she knew she and her son were two different people, so obviously their experiences would be different. Yet she could not let go of the timeline she had in her head, so when she lost her ability to speak sooner than her son, she grew very depressed, almost viewing it as a failure of some sort.

Again, my friend forgot that ALS experience$_1$ is not the same as ALS experience$_2$ no matter how closely people are connected. Hereditary illnesses may share similar diagnoses, but the lived experiences of the illness will necessarily be different. Had my friend been able to fully give up the timeline she somewhat forced upon herself, she might have felt less anxious and been able to focus instead on what accommodations she could make to help her cope with her own losses.

The following sets of statements first feature a verbal map that attempts to impose expectations from prior experiences and then a reframed statement that recognizes change happens.

Incorrect Verbal Map

"I just know I'll lose my hair again."

Reframed Statement

"This chemo treatment lead to my hair loss last time."
(The reframed statement leaves room for the possibility that this chemo treatment may result in either no hair loss or hair loss in a different way.)

Incorrect Verbal Map

"I shouldn't have to miss too much work."

Reframed Statement

"I hope to keep working, but I'll look into child care options just in case."

(This statement leaves room for the possibility that the new treatment regimen may be more tiring than expected or cause other complications.)

Incorrect Verbal Map

"Tom won't want visitors while he's at the hospital."

Reframed Statement

"Last time Tom didn't want people to visit him at the hospital."

(This statement leaves room for the possibility Tom might actually want visitors this time.)

In summary, no matter how familiar a situation may seem, we must try to recognize that no two experiences are identical, especially when encountering a negative life event. No one likes bad surprises, especially if they remind us of similar unfortunate experiences.

But a bad experience does not necessitate a second bad experience if we learn to adjust our maps and find new routes and destinations. Doing so is certainly an uncomfortable experience, especially when dealing with serious illness. However, being open to differences among seemingly similar illness experiences can pave the way for optimism and hope.

The following questions can help us use the indexing technique:

- What is different about this situation from other situations that seem similar?
- How would I describe how I have changed physically/emotionally/spiritually from the last time I experienced something similar to this situation?
- What do I anticipate from this situation? To what extent are my expectations based on past experiences? Is that appropriate?
- Am I comparing my own experience with someone else's experience? Is that appropriate?
- What don't I know about this situation? How can I find more information?

Change happens.

It's no secret our interests and tastes change over time. I used to love going to the mall in my 20's, but now I prefer to shop online. I didn't really exercise when I was growing up, but now I find myself trying to "keep things in place" on a weekly basis. Even my eating habits have changed. I used to eat a fairly large dinner every night while raising my son. Now that he lives on his own, I'm perfectly content with a bowl of popcorn or a waffle.

Notice how these reports of change feature words that "time stamp" my observations; words like "in my 20s," "now," "when I was growing up," "while raising my son." My use of time-stamping is hardly earth-shaking, but the words and phrases do subtly remind me and whoever is listening that things change over time.

Why is that important?

Because when we purposely talk in ways that suggest change could happen in the future, we can help reduce any stress or anxiety we might experience when those changes actually happen.

By using time-stamped qualifiers such as "often," "sometimes," "rarely," etc., our conversations create room up front for change to happen later on.

Think of a tradition your family maintains related to a major holiday. For example, perhaps you go to your grandmother's house for Thanksgiving, and you bring the pumpkin pies for dessert.

Now imagine you manage to break your arm and are prevented from making your famous pumpkin pie for the upcoming Thanksgiving feast. Certainly the injury itself sets the stage for frustration. However, the fact that you *said* "I bring the pumpkin pies for dessert" further exacerbates the stress you experience. After all, that's what you *do*.

But if you had reframed that statement years ago by inserting the word "usually" (saying "I *usually* bring the pumpkin pies for dessert"), room would have been created for other possibilities to happen right from the start.

And later, when your unfortunate accident actually occurred, the level of stress you experienced in not baking the pies would likely be reduced. "Usually" is not "always," so in essence, you gave yourself a free verbal pass for the year you broke your arm.

Our use of time-stamping in our conversations with one another creates room up front for change to happen later on. It alerts me and anyone else that adjustments can and may be featured along the way, and that's okay. Yes, we have a plan we hope to stick to that makes sense. But if things change, we will still be able to achieve our goals.

Illness journeys are no different. Anyone facing a serious illness will tell you time hardly moves in a chronological or predictable fashion. There are crazy ups and downs along the way that thwart any attempt to create a routine schedule. The key then is to time-stamp what we say in the first place to better cushion our needs to adjust later on.

Contrast the sets of statements below:

General statements

"I'm better."
"I'm nauseous."
"I'm cancer free."

Time-Stamped statement

"*Today* I feel better."
"I *often* feel nauseous for the *first few hours* after a treatment."
"I have not been diagnosed with cancer *for the past seven years*."

By adding a dimension of time to the above statements, room for change is created which, in turn, can help prevent us from being overly surprised or disappointed. Tomorrow I might not feel better, but perhaps there will be a time when I do not feel nauseous for quite so long after a chemo treatment. And if I were to receive a diagnosis saying some form of cancer had returned, my prior use of the above time-stamped statement could help remind me how long I have lived without cancer and also point to the possibility of living without cancer again in the future.

Experiencing changes over time can be uncomfortable, disappointing, frustrating, etc. But by using time-stamping in the first place, those negative feelings can be lessened to some extent and create small pockets of hope for better times in the future.

Another danger zone related to time-stamping is our

use of the words "always" and "never" in our conversations. These two small words can stir up strong emotional reactions because they position what we say in a "period, end of story" fashion. As a result, they shut the doors on any potential for change in what we are discussing, be it positive or negative.

Look at Sharon's husband's last comment in this chapter's opening conversation: "Yeah. We *always* seem to get bad news."

For those of us not familiar with Sharon and her husband, the statement really makes them appear as down-on-their luck people. We can imagine all kinds of failed hopes and dreams over the course of their lives by virtue of his use of the word "always."

But if we revisit the whole conversation, we learn Sharon actually experienced remission earlier in her life. How could that earlier report of remission have been perceived as bad news?

And what about their lives in general, outside of Sharon's cancer experience? Did they really *always* receive bad news? What about when Sharon and her husband found out they would be parents (assuming that happened)? Or what about the time her husband heard he was being promoted (assuming that happened)? Did they get a chance to experience some wonderful vacations together over the course of their marriage?

There were likely many times in their lives when each received positive news and life was good, yet the husband's use of the word "always" unfortunately erases those moments from being recognized or remembered.

The words "always" and "never" are both inaccurate and downright dangerous. They create false assumptions that, in turn, can make situations worse than they might otherwise be. To that end, we must be especially careful to avoid use of those terms in illness settings and better say what we mean. Look at the following examples:

"Always" statement

"He was *always* so positive before he had to go on dialysis."

> (This statement ignores the fact there were likely times before dialysis in which the patient was *not* feeling positive. Everyone has a bad day now and then. Perhaps the speaker just never witnessed those times.
>
> The statement also suggests the patient is no longer positive toward life in general. Again, the speaker just may not have witnessed times in which the patient *has* been positive while undergoing dialysis.)

Reframed statement

"Before he started his dialysis treatment, he seemed to be more upbeat when we would talk on the phone."

> (The revised statement is more specific about the observation being made—related to phone conversations—and leaves room for the speaker to be mistaken by using the word "seemed." It also creates opportunity for being pleasantly surprised if a future phone conversation *is* more upbeat.)

"Always" statement

"I *always* go with her to her appointments."

> (This statement sets the stage for negative feelings for both the caregiver and patient. For example, there may be times when the patient does not want the caregiver to join her at an appointment, but the

statement has created a precedent that prevents her from comfortably asking her caregiver to not attend. She fears he will feel insulted.

The statement also sets up the caregiver. He might not feel comfortable going to a particular type of appointment, but his prior use of the "always statement" prevents him from asking to not attend for fear of being viewed as being unhelpful or otherwise abandoning the patient. After all, he *said* he always goes to her appointments.)

Reframed statement

"I try to go to her appointments whenever she asks."

(This statement gives both the speaker and patient room to make exceptions along the way.)

"Never" statement

"We would *never* put mom in a nursing home."

(While well meant, this statement locks family members into a situation that may grow beyond their control. For example, if their mother faces unanticipated dementia challenges, it may not be possible to allow her to live on her own or with family members. Having made the "never" statement, family members can experience incredible guilt if they later do need to request nursing home services.)

Reframed statement

"We will try to provide the best care for mom as we are able."

(This statement gives the family members latitude to pursue all types of care with loving intentions and desires.)

For most of us, learning to time-stamp what we say in our conversations will take practice. But doing so is well worth the effort to help prevent us from locking ourselves into static mindsets. Change *does* happen, and our language choices need to create room for that to happen.

Here are good questions to ask ourselves related to time-stamping our conversations:

- Does my report feature time-based qualifiers such as the words "sometimes," "often," "usually," etc.?
- Am I avoiding use of the words "always" and "never?"
- Am I adjusting my reports over time in ways that reflect changes I observe?

The perils of inferences and judgments

As mentioned throughout this book, illness is fraught with uncertainty. Even if a diagnosis seems unquestionable or is similar to that which we experienced before, each illness journey is unique and finds us wondering what will happen next.

Key ways in which patients and caregivers attempt to reduce that uncertainty is to sprinkle their conversations with inferences and judgments. They attempt to not only predict what will happen based upon prior knowledge or experience (inferences), but they also attempt to evaluate progress along the way (judgments).

These types of reporting make sense in terms of allowing patients and caregivers opportunity to reflect upon their illness journeys. However, the danger in relying upon inferences and judgments lies in our tendency to view those types of information as "factual"—reports that can be proven or measured.

In reality, inferences and judgments are more akin to hypotheses and best guesses than they are to authentic logic or science. There is ample room to be mistaken with inferences and judgments, so we need to avoid using either if at all possible. Not an easy thing to do!

Again, an "inference" is a statement made about the unknown based upon the known. When we infer, we attempt to apply the limited understandings and experiences we have gained over time ("the known") to newer, unfamiliar situations ("the unknown") as a way to reduce the uncertainties of those new situations.

The challenge in doing that? "What we know" often does not fully fit new situations. Sometimes our inferences can be closer than not, and when that happens, we experience lower levels of stress (an "oh phew!" type of experience). But when we rely heavily on inferences that fail to come true, we experience genuine disappointment, frustration, and other negative emotions.

Pretend we have been good friends for a long time, and knowing I teach communications, you ask me if I think your 19-year-old daughter would be a good candidate for our program. Any response I would offer would necessarily be an inference based upon what I observed and learned about your daughter over time (i.e., her ability to speak and write well, her comfort in talking with diverse people, reports from you about her academic skills, etc.).

Despite both of our knowing I was sharing my best guess (inference), we would still tend to perceive what I said as factual since your daughter is *known* to me and I also *know* about our program.

But can I really know if your daughter would succeed in our program based upon what I know of her in general? And can I really know what it is like to be a student in our program, since I have only experienced it as a professor?

Possibly yes to both questions, but there is definitely room for my being mistaken. And should I be wrong, unfortunately for both of us, our having interpreted my inference as "fact" sets us both up for frustration, disappointment, etc. I end up feeling embarrassed if my predictions do not come true, and you feel upset that I wasted your money and your daughter's time.

"Judgments," which are equally dangerous, are statements (often inferences) that include evaluative elements such as good or bad, right or wrong, smart or stupid, etc.

Like inferences, we tend to view judgments as facts since the evaluative elements are often emotionally charged. The judgmental word choices literally get a rise out of us that prompts us to invest a higher level of energy into the idea being communicated.

Think about this statement: "He flirts when his wife is not around." Is this report something that can be proven (fact?) We would first have to define what behaviors constitute flirting, and then we would have to be with the gentleman at all times when his wife was not present to see how consistently those behaviors occurred. The squishiness of the word "flirts" combined with our inability to be with the gentleman at all times prevents the statement from standing as fact. Instead, the report is an inference laced with judgment. How so?

First, the inference comes about when the speaker applies her own interpretation of multiple conversations she had with the gentleman when his wife was not present (the known part) to all other experiences in which the gentlemen speaks to other women when his wife is not present (the unknown part). She then presents that information as fact by not including any qualifiers in her statement that leave room for her to be mistaken (i.e., introducing the statement with "in my experience" or "based on conversations I have had with him"). She simply reports the man is a flirt.

The judgment presents itself through her choice of words to describe the gentleman's behavior. Based upon our cultural norms, we generally perceive married men who flirt as unsavory louts. And because that norm is so ingrained in our way of living, we easily perceive her judgment of the gentleman's character as being factual. After all, what type of husband flirts with other women? Think back to your initial physical and emotional reactions to the statement. Did you feel at all sympathetic toward the man? Did you question whether or not the statement was even true?

To make matters worse, the judgmental nature of the speaker's statement will necessarily bias the way we will perceive the man if we ever meet him and will influence what we say and how we behave toward him as a result.

If we are a friend of the speaker and have no relationship with the gentleman, we will be inclined to interpret her report as fact and greet him with suspicion. If we are female, we will be especially alert to anticipated behaviors based on what the speaker reported. (This phenomenon is at the root of most stereotypes.)

Now look at the statement again. Couldn't it be the guy just tries to be engaging and he actually adores his wife?

Absolutely. But our knee-jerk semantic reaction to the speaker's judgmental report unfortunately makes it easier for us to leap to the more negative conclusion and consider the remark as factual. This is the danger that exists with inferences and judgments and is why they should be avoided whenever possible. Inferences and judgments are limited in scope and subject to change once new information is introduced.

In terms of illness settings, our temptation to use inferences and judgments in our conversations increases in direct proportion to our levels of uncertainty. Illness often comes crashing into our worlds, and we find ourselves forced to identify best next steps while still trying to catch our breath from the initial blow of the diagnosis.

It is likely the illness is something we never experienced before, so we have no prior knowledge to help us navigate the new territory. Even if we had tried to imagine what we would do if ever faced with a scary diagnosis, it is likely the actual lived experience will be nothing close to what we thought would happen.

So we try to apply what we think worked in the past for seemingly similar situations (the "known") to the vast uncertainty that lies ahead (the "unknown"). Such inferences are meant to bring us comfort and a sense of control, but they can easily backfire.

For example, I often hear spouses of those diagnosed with a life-limiting illness offer comments such as "knowing her, she'll bounce right back" or "he's never been a quitter before, so I doubt he'll start now."

While each of these inferences is meant to be encouraging and could come true, each also sets up the patients for future stress if the inferences do not come true. Anyone who has experienced chemotherapy or other difficult treatment regimens knows it is difficult to "bounce back" or to not feel like quitting at times, so not being able to fulfill loved ones' voiced predictions of being able to rise to the occasion can result in a variety of negative emotions when that doesn't actually happen.

This is also the case when a patient draws similar positive inferences about herself that cannot be maintained no matter how hard she tries.

Notions of "failure" — a hugely judgmental word — are

introduced when inferences are not fulfilled. We want to be viewed as emotionally and physically strong. We do not want to be viewed as complaining or weak. So think of the patient who hears her husband tell a neighbor "she's a fighter...I just know she'll beat it." It is likely she will feel compelled to achieve that image her husband has of her, which prevents her from complaining or considering making different choices, including foregoing treatment and making the best of the time she has remaining.

One of the greatest ways we can demonstrate empathy toward loved ones who are hurting is to try to pull together words of encouragement and support. And often we do so in ways that suggest our loved ones already have what it takes to succeed based on what we know about them (inferences). Patients do the same for themselves to self-motivate.

While I would never suggest such hope-related inferences be completely stricken from our conversations, I would caution us to at least develop the habit of adding "qualifiers" to those reports to remind us there is room for the possibility we are mistaken or not fully informed. By doing so, we can help reduce future disappointment if there does come a time when our inferences are incorrect.

Below are two sets of inferences. Note how the first set presents information in a factual manner and how each inference sets the speaker and the subject up for disappointment should the inference be contradicted in the future.

In contrast, the second set of inferences features qualifying statements that create room for change to happen within the "unknown" part of the inference.

Inferences Without Qualifiers

> "I don't expect any real problems, since I haven't had trouble with dialysis over the past year."

"She beat the odds last time, so I don't see why this should be any different."

"Everything I've read suggests we are ahead of the game since we received the diagnosis early on."

Inferences With Qualifiers

"I didn't seem to have trouble with dialysis over the past year, but I know things can change over time."

"She apparently beat the odds last time, but we still plan on taking each day at a time."

"Everything I've read suggests his early diagnosis bodes well for recovery. That would be wonderful if that happened, but we also know every case is different."

The process of adding qualifiers to inferences is not meant in any way to be a downer or reduce hope. We need to envision opportunities for healing to keep ourselves going. Adding qualifiers to our inferences merely serves as a reminder to ourselves that things change over time, and they also help cushion the blow to some extent when our inferences prove to be incorrect.

While hope-filled inferences can be shared if modified with qualifiers, less-than-hopeful inferences should be avoided completely. Reports like "I always seem to get bad news, so why should today be different?" only add to existing stress, and there is no way to successfully modify a negative inference with qualifiers. When a patient or caregiver infers one is likely doomed to repeat negative experiences from the past, they do not enhance or promote productive conversations.

Even when it seems like things always go wrong, the

fact is things change, and there is always a possibility for something positive to emerge at some point in time (the "unknown"). In this regard, remember what our mothers told us: if you don't have anything nice to say, don't say anything at all.

Examples of negative inferences to avoid altogether

"Our family has a history of heart disease, so why should I be any different?"

"Every drug I've taken has made me sick, so I'm betting this treatment can't be any better."

"She's always been a smoker, so what makes you think she'll quit now?"

In addition to trying to avoid negative inferences, statements of judgment should also be eliminated from our conversations as much as possible. Judging ourselves and others often leads to either/or orientations (good/bad, right/wrong, success/failure, etc.) that as noted earlier are inaccurate and inappropriate.

Once we judge a situation or person, it is difficult to adjust future reports about that situation or person that do not support our original judgments. We lock ourselves into initial, often premature, judgments that prevent us from thinking in new ways. After all, who wants to admit to being wrong?

Within the world of illness, judgment often emerges in two key types of discussions: what caused the illness and how the patient and caregiver respond to the illness.

Judgments related to the causes of illness can be especially destructive. While there are behaviors and genetic factors that can significantly contribute to a difficult diagnosis, oftentimes one cannot be certain a particular aspect of how one lived one's life is the absolute cause of an illness. Therefore, trying to place blame on a specific behavior may not only be inaccurate but is also unproductive at the time of diagnosis.

Examples of judgment statements related to illness causes to avoid altogether

"He never watched what he ate." (diabetes diagnosis)

"She knew there was a history in her family, but she didn't get regular checkups." (breast cancer diagnosis)

"He was an alcoholic." (liver disease diagnosis)

The judgmental nature of each of the above statements sets the bar for subsequent conversations and behaviors toward the speaker and the patient. Each statement's negative overtone suggests the patient somehow failed to do what was right and somehow deserved the diagnosis. Furthermore, the speaker's use of a judgment statement positions her as being better than or somehow more knowing than the patient.

While our behaviors can certainly contribute to the onset of an illness, we must remember for many illnesses there is rarely a single event or behavior that results in a diagnosis.

Furthermore, time spent trying to imagine what could have been done differently will not necessarily lead to recovery. And suggesting behaviors were right or wrong, good or bad, gnaws away at a patient's self-esteem. None of us can change the past, we can only move forward.

Judgment statements related to how a patient or caregiver manages the illness process can be just as damaging to self-esteem and relationships. Notions of "best practices" and "best care" do exist within the medical arena, and they necessarily introduce a level of judgment on the part of medical practitioners, patients, and their circles of support.

Certainly such judgments are important to insure care is appropriate and warranted, especially when it comes to prescribing treatment regimens. After all, we pay doctors and specialists to provide their best judgments about what steps should be taken.

But when it comes to our daily conversations with one another outside of the doctor's office, judgmental statements about how patients and caregivers are managing their illness journeys can prove damaging. Caregivers can be especially judgmental of themselves as they struggle to accommodate and support newfound challenges their loved ones face. They can get caught up worrying if they are offering help in the "right" way or possibly making things worse.

While I would not suggest completely avoiding such reflective conversations, I would encourage caregivers to work hard at "reframing" judgments into statements that solely report observed behaviors or feelings. The goal is to eliminate words that suggest there is a best or better way of doing things. Instead, caregivers should simply describe what happens along the way and then imagine possibilities in light of those observations.

Original Judgment Report

"I really thought I could handle helping her in the bathroom."

(The statement suggests the speaker is incapable of achieving expectations he set for himself which results in an overall sense of failure.)

Reframed Report

"I feel awkward helping her after she uses the toilet."

> (The revised statement focuses on a particular activity that the speaker finds difficult. It helps the speaker recognize there are other bathroom activities he handles well, such as showering and washing up, and it creates room for exploring options to reduce the awkwardness he feels with this particular task.)

Original Judgment Report

"I let him down when I cried at the doctor's office."

> (The statement suggests the speaker's act of crying made her appear less strong in the eyes of her husband, and as a result, he was disappointed in her.)

Reframed Report

"I cried in the doctor's office."

> (The revised statement simply reports what happened and doesn't attempt to judge the behavior in regard to her relationship with her husband.)

Original Judgment Report

"I'm not smart enough to keep track of all the medications and what they are for."

> (The statement suggests the speaker is totally unable

to understand and monitor a patient's medication schedule.)

Reframed Report

"I need to create a system to help me track all her medications."

(The revised statement doesn't question the intelligence of the speaker and offers a proactive behavior to rectify the challenge she faces.)

The same act of reframing judgments can be employed by a patient as well. For example, patients often judge themselves in direct proportion to the amount of help they need from others. The more self-reliant a person was before her illness, the more judgmental she seems to be in needing assistance from others.

Phrases such as "It's like I'm a baby again" or "I can't even wash my own face" are hugely judgmental, and more importantly, they are not true. It is simply impossible for an adult to turn into a baby, and equating the need for help as being baby-like is inaccurate and inappropriate. People need help in all sorts of ways at all ages. And how does not being able to wash one's face make one a lesser person?

Many of us fear having to receive, let alone request, support from others, as we view it as a sign of weakness or dependence. We do not want to be obligated to others, and we sure as heck do not want to appear vulnerable.

However, judging ourselves negatively in light of the help we need from others is harmful to both ourselves and those who try to offer us support. We need to turn our negative self-judgments on their head by using different words to describe the help we receive (reframing).

Original Judgment	Reframed Phrase
I'm helpless.	I need help walking and bathing.
I'm dependent.	I rely upon the support of others at times.
I'm unable to do that myself.	You make that easier for me.
I can't even drive.	I use the center's shuttle service.

We need to examine our language choices when we describe what we can and cannot accomplish on our own. An illness can certainly make it difficult or impossible to manage all that we managed on our own prior to the illness. But isn't that true for other things in life? Haven't we all experienced times in which we needed help from others?

Needing help with more intimate activities, especially those relating to bodily functions, can be extremely difficult to request or accept. But with use of reframing, the stress can be reduced a bit.

In summary, inferences and judgments are best avoided whenever possible. They do not offer accurate predictions or descriptions, which means they do not reflect the territory, and as a result, lead to disheartening conversations in our own heads and with others.

If we cannot avoid offering an inference or judgment, we need to add qualifiers and reframe what we say to better tell it like it is. To get us there, we need to challenge what we interpret as "factual" by asking ourselves the following questions:

- Am I basing what I say on prior experience? If so, does that experience accurately relate to this new situation?
- What is my source(s) for speaking? Is that source appropriate?
- Is what I say really factual? Can it be proven somehow? If not, am I sharing an inference and/or a judgment?
- Do my word choices place blame or point fingers at myself or others? Such reports should be avoided if possible.

Chapter 4

The Map Reflects the Mapmaker

"We see the world as 'we are,' not as 'it is,' because it's the 'eye' behind the 'I' that does the seeing."
--Anais Nin

A note before we begin this chapter...

Most patients recognize that use of prescription drugs, radiation, and other illness treatment regimens can influence the functioning of their brains and overall nervous systems. For example, cancer patients often describe experiencing "chemo brain" that seems to affect their short-term memory and causes confusion, disorganization, mental fatigue, etc.

The Mayo Clinic has reported that although the diagnosis of chemo brain has not been formally confirmed (researchers are not entirely clear which drugs specifically cause which symptoms), there does seem to be a shared set of neurological symptoms many cancer patients experience with chemotherapy.

In addition to medications or other treatments, tumors or other physical evidence of an illness can also affect the nervous system by releasing substances into the bloodstream and/or by physically imposing upon the nervous system itself. Such cases can also affect one's ability to think and reason, as well as one's five senses.

In light of these concerns, patients experiencing medical treatments or other challenges that seem to be affecting their cognitive abilities should not feel obliged to apply the strategies featured in this chapter. Doing so could lead to frustration, disappointment, and other negative feelings, especially when cognitive changes are out of one's control.

However, I do still encourage patients and caregivers in those situations to rely upon "the word is not the thing" and "map is not the territory" strategies when discussing the changes they are experiencing.

Losing one's ability to think and reason to any degree can be both frightening and embarrassing. Often patients will make comments like "I'm losing my mind," "I'm not myself," or "I must be going crazy."

None of those statements is entirely true, and they serve only to increase levels of stress and anxiety. One cannot literally lose one's mind, and the fact one is still able to communicate and manage some aspects of one's life suggests the nervous system is doing the best it can.

To suggest one is "not one's self" is also untrue. One is always one's self, but that self is in constant flux. Changes happen to our bodies and minds daily, with and without introduction of medications.

And suggesting one is "going crazy" is untrue and unhelpful. Increased instances of forgetfulness and an inability to always find the right word does not necessarily mean one is losing one's touch with reality.

I would also encourage patients and caregivers to be patient with one another in light of any cognitive changes that occur. Active listening skills such as paraphrasing and asking for clarification along the way can help meaningful conversations continue.

Finally, patients and caregivers should check with their doctors to explore possible side effects to the nervous symptom of any treatment regimen. Attention is often paid to anticipating the physical side effects of treatments (nausea, hair loss, etc.), but patients and caregivers should also explore the possible side effects on the brain and nervous system.

It starts with the brain.

The verbal maps we create each day directly reflect the mechanics of our brains—the unique ways in which our brains manage electro-chemical impulses in relation to different stimuli. Different regions of our brain work together to help us perceive our environment, and then we attempt to translate our perceptions into a language that allows us to examine, discuss, sift, censor, etc. those perceptions. Simply put, we speak as we *think*.

To set context, let us first examine how the brain functions. The brain features a variety of regions that correspond to different types of sensory functions, reasoning abilities, physical movement, etc. All of those regions of the brain, combined with the spinal cord, make up the human nervous system.

That system houses millions and millions of microscopic neuron cells that help the nervous system receive and send information when we encounter any type of outside stimuli. Each neuron has "branches" (dendrites and axons) that help it connect to other neurons to share that sensory information. When two neurons connect, it's called a "synapse."

While we are born with all the neurons we will ever have, most of them are not connected at birth. Thus, our physical, emotional, and intellectual abilities are quite limited. But as we develop and encounter new stimuli, information connections (synapses) occur that help us perceive and process those new experiences.

When our brain's neurons first connect in response to a particular event or situation, that connection creates a small "pathway" on which other information related to that particular event or situation will travel in our brain. The neurons' branches join together to create a sort of primitive "road" on which information related to that situation or event flows.

If we continue to repeatedly experience that same situation or event, the pathway becomes ever more established in our brain—the branches bond more strongly and we learn that information. If the situation or event is repeated enough over time, we eventually create a "superhighway" in our brain on which that information flows more quickly and easily. The more often our brain sends electrical messages down the same path, the stronger that path becomes.

Think back to when you first learned to tie your shoes.

Initially you were likely all thumbs as you struggled with wrapping and weaving the laces together into some semblance of a bow. That difficulty stemmed from the fact your brain didn't have any neuron connections related to tying shoes. There was no "shoe tying pathway" in your brain. Instead, you had to create that connection through continued practice until the connection (the pathway) became more familiar.

Now after years of tying your shoes, chances are your brain now features a sort of shoelace superhighway that allows you to quickly and easily accomplish the task with minimal thought or attention.

The verbal maps on which we base our thought and actions come to life in a similar fashion. For example, for most of my school-aged life, I developed the verbal map that one needed a college education to land a good job. That neural connection in my brain (idea) came about through repetition of that information from multiple sources. My parents and grandparents not only introduced the idea (they planted the first neural connection in my head), but they kept repeating it throughout my elementary and high school years. My teachers also told me college was critical for employment, TV commercials for area colleges and universities told me I needed a degree for success, and my friends (whose parents told *them* college was important) told me we had to go to college. Even family board games such as "Life" and "Careers" suggested a college degree was critical for success.

As a result, by the time I reached the end of my senior year in high school, I had one heck of a neurological superhighway in my brain that said "college or bust!"

When the time to apply to college actually arrived, though, I realized I really had no desire to go. Instead, I simply wanted to get a decent paying job that would allow me to move into my own apartment.

Needless to say, it was a real struggle for me and my

family to get off that college-bound verbal superhighway. After all, that verbal roadway was wide open...the way had been cleared and paved through report after report after report. Why would I choose then to veer off at some unknown exit?

It was both frightening and frustrating for all involved, but ultimately I attended a two-year secretarial school and managed to piece together a fairly successful existence.

The older we get, the more superhighways we develop in our brains. That is not all bad. Our brains' superhighways allow us to process information more quickly which, in turn, leads to efficiency and predictability of thought.

However, the disadvantage of our brains' superhighways is those embedded thought patterns turn into personal habits and paradigms that are hugely difficult to shake. The stronger the neurological connection in our brain, the more difficult it is to revise or eliminate and the more stress and anxiety we experience in attempting to do so.

Furthermore, the superhighways we develop in our brains can actually prevent us from learning new or alternative ways of thinking. Research shows we literally are unable to consider new ideas, let alone generate new ideas, if we have convinced ourselves that what we know is true.

For example, if I convince myself I do not look good in the color yellow (I build a no-yellow-fashion highway in my brain), it is possible for me to go into a department store and literally not notice yellow garments hanging on the rack. In my brain, yellow garments are simply not an option, and so even if they exist, my brain will simply overlook them.

This subliminal oversight is one of the ways my brain tries to protect me from any uncomfortable feelings I might face if I even attempted to consider wearing yellow. That can be a good thing if yellow indeed makes me look terrible. However, if the color actually enhances my appearance, I'll lose out because I cannot get off the superhighway I created in my brain—I essentially talked myself out of the opportunity to look my best.

An example within an illness setting could relate to the onset of Parkinson's disease. Most of us travel along a verbal map that says Parkinson's is an old person's illness. We've read that is so, we've heard that is so, and most of us have only witnessed Parkinson's in older people.

So when actor Michael J. Fox announced he had Parkinson's in his 30's, we were stunned. The road we had been traveling now featured a new exit most of us had not imagined possible.

What if Mr. Fox had clung to the idea that only old people suffered from Parkinson's disease? What if he ignored all the signs and refused to entertain the possible diagnosis?

I do not know for sure how Mr. Fox initially reacted to the diagnosis or even pursued it in the first place, but fortunately at some point he constructed a new verbal "Parkinson's highway" in his brain that has enabled him to meet the challenge and become one of the greatest advocates for a global cure. He recognized the verbal map we had all been applying to the illness was inaccurate, and we would only go nowhere fast if we didn't get off that neural highway.

Like Mr. Fox, when we experience stress or anxiety in life, we need to examine what road we are on in the first place. Is the stress coming from our clinging to a verbal highway that simply doesn't get us where we want to go? Should we think in new ways (take a different road)? Are we banging our own heads against a wall?

While the uncertainties we face in venturing off the beaten path can appear to be more stressful than the actual problem we face, if we allow ourselves to imagine and create new maps, we can redirect ourselves in more positive directions. We can literally create new neurological connections in our brains that allow us to become more flexible and creative in our outlooks on life.

How to do that? We need to set ourselves up to receive new information and have new experiences on an ongoing basis. We need to seek out the unfamiliar in ways that force our brains to literally be caught off guard. We need to enter new situations without expectations and an open mind.

Obviously, this is so much easier said than done, especially within an illness setting. How can one possibly maintain an open mind toward heart disease or liver failure?

In Chapter 3, I described the possibilities of finding positive meanings within negative life events. Our brains play the lead role in allowing that to happen, as they are the only mechanisms we have that can help us create new roads and routes to discover those meanings.

As painful and scary as it may seem, we need to remember our brains *are* capable of helping us imagine the positive that exists within dire situations. At first, we will likely find ourselves bushwhacking through jungle-like territories. But if we keep at it, eventually we will start to clear some new trails in our brains (neurons will connect in new ways), which might later become more apparent and natural roads of thinking.

We must also remember that because of the unique ways in which our brains function, our maps can never be fully identical to another person's maps. Everyone's neurons connect differently, and while there are biological similarities in terms of the process of mapping in our brain, the actual pathways and superhighways we develop within our brains are unique to each of us.

Even when we seem to agree, the actual thinking process that occurs in our brains is still different from others' thinking processes. I might take a superhighway to get to our shared thought, while you take two surface roads. Still, because we agree, we do not think about the differences in *how* our brains led us to those conclusions.

On the other hand, when our ideas collide, we are forced to recognize our brains do function in unique ways. How often have you heard or used the phrase "what were you thinking?" Our frustrations, disappointments, etc. suggest others' brains are not working in the "right" way—which essentially means "our" way.

In these situations, we need to remind ourselves it is impossible for any of us to think in the exact same fashion. Even if we were able to eventually persuade the other person to "think like us," our neurological connections would still necessarily be different, since the ways in which the connections were developed came about differently. My neurological connection may have come about through direct experience over a long period of time, while your neurological connection may have come about from reading an article or watching a TV show. As a result, my connection is more of a superhighway, while your connection is more of a fledgling graveled road, since the information is new to you. You might even change your mind and disagree with me again, since the neurological path is not as ingrained in your brain as in mine.

When our maps do collide, we must resist our knee-jerk attempt to persuade the other or otherwise prove what we think is the right way to think. Instead, we should value our diverse electro-chemical ways of thinking and understand those differences can create room for growth and lead to construction of new roads in our own maps.

We should think of the act of considering others' viewpoints as a form of exercise for our brains in which we stretch and expand our neurological connections. Doing so is good for us, as new connections in our brain enable us to become more flexible in our thinking. We can imagine new possibilities and ways of behaving, which, in turn, help to reduce stress, anxiety, and other negative feelings.

The following questions can help us better develop and strengthen the neurological connections featured in our verbal maps:

- How am I looking at this situation? Is there a different way to view it?
- What does my point of view overlook or not take into account?
- Where does my reaction come from? When and how did I learn what I "know?" Is that source still reliable? appropriate?
- What do other people or sources say about this situation?

Sensing a difference

Our five senses (sight, smell, touch, taste, and hearing) also affect the way we perceive and then interpret our world, and they are as unique to us as individuals as are our brains. While we are eager to describe and explain our sensorial experiences to others, our language is often hardly up to the task.

What word best describes the scent of a rose? What word suffices for the sweetness of a ripe nectarine? How would you describe the smoothness of silk? See the difficulty? Our senses are often far more heightened than the words we try to use to describe them.

To compound the challenge in discussing what we sense, we so easily forget others do not experience the world in the exact same way. For example, our ability to see varies across the population. I can read signs from a farther distance than can my husband. I'm also able to distinguish between black and midnight blue with little effort. But I have to admit trying to thread the needle on my sewing machine is getting more difficult these days!

Now think about your sense of taste. Do you enjoy spicy foods? Sour? I have difficulty with both, so it is likely your report of eating a jalapeno pepper or sucking on a lime will be far different from my own.

How is your hearing? Can you sleep through crashing thunderstorms, or do you hear even the slightest sounds? Do you listen to and enjoy the same music you did 10 years ago? What sounds do you love? What sounds do you detest?

Just as we say what we *think*, we also say what we *sense*, and our sensitivities are as electro-chemically diverse as are our abilities to think and reason. One step to accommodate for that difference in our conversations is to adopt a "to me" attitude when we attempt to describe what we sense:

To me, that paint color is just right for the kitchen.
Sushi tastes fishy to me.
Wool is a scratchy material to me.

Qualifying our statements as relating only to ourselves helps prevent us from assuming others feel the same way we do. It also helps reduce possibilities for judgment and offense in our conversations.

Imagine having dinner with your best friend who carefully prepared a vegetable you have never enjoyed eating. If you were to announce "this [vegetable] tastes bitter," chances are your friend would feel judged and likely take some offense. If, however, you explained "this [vegetable] can taste bitter to me," the friend might feel a bit less dejected at your passing on her side dish.

In terms of an illness scenario, issues related to odors often arise, especially in hospitals and nursing care facilities. Visitors can be overwhelmed by various smells when they visit loved ones on a sporadic basis. On the other hand, patients who stay for longer periods of time may not notice the odors as much. The visitors need to remind themselves their senses of smell are different from the patients' senses of smell, and there is not necessarily a right or wrong to the matter.

In addition to our senses being unique to us, our senses also change over time. That which we loved to eat can suddenly become less tasty. That which we were able to easily see can become more blurred. That which we could hear now sounds a bit muffled.

These changes mean our verbal maps must change as we talk with others. The map I presented 15 years ago related to my keen ability to thread a needle without difficulty does not fit me anymore. Nor does the map about my love of a certain flavor of soft drink. That mapmaker (myself$_{then}$) no longer exists, so a new map has to be crafted by this newer mapmaker (myself$_{now}$).

Consequences of illness upon our senses often force our need to recognize we are changed mapmakers. That which we could sense before is not the same, just as that which we never sensed before is now present. While those changes can be frustrating, surprising, or any other sort of descriptor, the fact is the maps must now change to reflect the new mapmaker.

An example could be a gentleman who always enjoyed his wife's cherry pies, but after starting a new medication, they just do not taste the same to him. He does not enjoy sweet flavors as much as he used to, and in fact, he'd prefer cheese and crackers if he had a choice.

Now imagine the wife's reaction to this change in her husband. Chances are she will at least feel disappointed, and she might even take his change in taste as somehow related to his feelings about *her*. After all, he *always* liked her cherry pies.

Both the gentleman and his wife need to remember our senses change over time, and that is okay. All we need do is adjust our maps accordingly and try not to interpret those changes as anything more than a change in one's sense of taste, touch, sight, hearing, or smell.

Taking time to notice

Imagine you have the opportunity to view the beautiful gardens of Italy's Tuscany region. Which mode of transportation would provide you the best sightseeing opportunity: passing by the gardens on a speeding train, riding a bicycle by the gardens, or actually strolling through the gardens?

Most would say a walk through the gardens would afford the best opportunity, since we would be moving more slowly and taking more time. However, all three will actually result in limited observations.

First, in regard to our senses, there is an abundance of "energy" that exists in the gardens that we cannot even sense, such as ultraviolet light, photosynthesis, satellite signals, etc. Our senses are simply not able to notice such things without the aid of technology.

But as importantly, regardless of how slowly we view the gardens, we simply cannot see *everything* in the gardens...each leaf, each branch, each particular flower, each butterfly, etc. Instead, we find ourselves scanning the garden vistas overall, and after doing so, we might then choose to especially observe certain aspects of the gardens, such as the water fountain, the intricate blooms of some orchids, etc.

Choosing to look at particular aspects of the gardens (paying attention to certain details) necessarily forces us to overlook or miss other aspects of the gardens. We cannot see it all.

In addition, any form of distraction can further prevent us from observing all aspects of the gardens. For example, a member of the tour group might be overly talkative, so despite our best efforts to fully examine a lovely hidden pond, his incessant babbling eventually drives us away. Similarly, discovery of a certain lily we have never seen before might prevent us from attending to other flowers in the garden.

Not all distractions are external to us. Personal fatigue, for example, can prevent us from attending to a sight or experience, as well as personal interest or curiosity. That which interests us tends to get most of our attention. Routine or otherwise very familiar sights or experiences tend to garner less attention.

Our act of choosing where and when we pay attention (purposeful or not) necessarily forces us to ignore other aspects of the gardens. Even if we were to *crawl* through the gardens, our ability to attend to and notice all they hold would be limited, and as a result, our verbal maps for the gardens would be equally limited. Something would always be left out, either by choice or by our sheer inability to process everything.

Not being able to observe and know everything does not mean we cannot function or communicate in a healthy fashion. For example, my not being able to physically detect the microwaves in my oven as it heats up my leftover pizza will not make the pizza less tasty. Nor does my oak tree depend upon my being able to discern the process of photosynthesis for that process to occur. Much of what we miss going on in our environments does not necessarily hurt us.

There is a danger, however, when we forget we *do* miss much in our environment, and as a result, we deny the existence of that which we miss. We can be easily tempted to say that which we did not or could not observe directly either does not exist nor has nothing to do with us.

We can also set ourselves up for unpleasant surprises when we forget we cannot know everything about everything. This is especially true in illness settings where a scary diagnosis seemingly comes out of nowhere. Patients wonder how they could have missed the signs of illness or try to imagine what they could have done to receive such frightening news. They worry they did something wrong or had otherwise let themselves and their families down for not knowing what was going on with their bodies.

First, as noted above, patients need to remember their senses are quite limited. There was no way they could have seen, heard, touched, etc. the biological processes happening at a cellular level within their bodies prior to the diagnosis.

Second, they likely had no real reason to pay attention to the possibility of the illness developing. Even if they experienced symptoms they later learned were connected to the illness, chances are they were distracted by other things in life, so they didn't consider the possibility.

Finally, after receiving the diagnosis, patients forget their brains can only process limited amounts of information at a time, especially when information is new. If they focus their attention on the symptoms and causes of the illness, their attention to information on treatment options or finding the right doctor is somewhat reduced. Or if they focus on nutrition and exercise, they are unable to also fully discern the differences among treatment plans. Again, our brains can only process so much information at any one time.

Shifting attentions lead to shifting maps, and that is okay. What is not okay is purposely ignoring maps that we perceive to be more negative. When our attentions are tapped toward negative life events, it is unhealthy to not invest more of our attention in those matters. Ignoring a situation prevents us from gaining more information that can better inform the verbal maps that guide our thoughts and behaviors. As a result, voluntary inattention sets us up for unhappy surprises and an inability to generate new maps in a timely manner.

We need to be willing to pay attention to as much as

possible in our illness settings—the good and the bad—so that the verbal maps we create are as accurate as possible.

Certainly increased knowledge can lead to increased anxiety, however, that knowledge also arms us to become better mapmakers once we are able to organize new information in a meaningful fashion. Hopefully, our ability as mapmakers to put that increased knowledge into the proper perspective will ultimately help reduce the anxiety and create room for new possibilities and hope.

The following questions can help us be more mindful of the time and attention we have invested into our maps:

- Am I rushing to judgment?
- What distractions am I experiencing right now? What might be preventing me from thinking more critically?
- What don't I know about this situation or event? How can I gain that information?
- How purposeful was I in exploring this situation or event? What level of attention did I give to that situation or event? Was that appropriate?

Establishing new routines

The old adage "we are creatures of habit" not only rings true, but it also describes another brain function that can limit our take on reality. Neurological superhighways described earlier in this chapter not only limit the scope of our thinking, but they also establish personal habits that narrow our ability to behave in new ways.

A good example of this phenomenon can be found in our daily routines each morning. Like many people I know, I have a definite order of events when it comes to showering, styling my hair, and getting dressed each day. Hair gets washed before the body (so conditioner can work its magic), arms get dried before legs, deodorant goes on before styling products, pants go on before tops, etc.

Those habits of mine are actually neurological superhighways in my brain that relate to "morning routine," and trying to shuffle or adjust that routine is just as difficult as trying to adopt new ways of thinking. Just ask someone who breaks an arm or leg how uncomfortable and frustrating it is to try to maintain one's morning routine. We don't realize how ingrained our behaviors have become until they are seriously challenged.

To further complicate matters, we tend to identify our habits as being the right way to do things. This, too, relates to the physical discomfort we experience if we are unable to fulfill our habits.

At times I have tried to force myself to adjust my morning routine, and doing so did not feel good. Drying my legs before my arms physically felt uncomfortable and awkward. And I was so distracted putting my blouse on first, I actually fumbled a bit with the buttons.

The comfort we find in our behavioral habits is akin to that which we experience in our mental stereotypes and paradigms. They just feel "right" to us. Unfortunately, though, that act of ascribing judgment to our habits and routines makes them even more difficult to escape.

Illness does not care whether we tend to do laundry on Saturday evenings or we always get dressed before we eat breakfast. It has no concerns whether or not we make the bed each morning or tend to load the dishwasher in a particular way.

Illness does not seek to accommodate our routines and habits in any shape or form, and at times, it can seem as if illness is solely determined to turn our way of life on its head.

To help reduce the anxiety and stress of having our routines seemingly tossed out the window, we need to remember that habits and patterns of behavior are *neurological* reactions that can be rerouted and modified. Certainly there are levels of discomfort in forging new ways of doing what has been so familiar to us, but our brain is equipped to help us do that if we allow it to do so. As noted earlier, our brain is amazingly complex in its ability to establish new neurological connections if we create opportunities for it to do so. And once we remove judgment about doing things "the right way," the brain is freed even more to be creative and to identify new ways of managing our daily activities.

Maybe we can do the laundry more than once a week. Perhaps we can straighten the bed neatly enough without fully making it. Possibly we could eat a bowl of cereal before getting dressed. Maybe we could skip church now and then.

Illness usually leads to change at some level for all involved, and while that change is uncomfortable at first, it does not have to completely upset our worlds. Remember, the brain that got us into our habits and routines can also get us out if we let it.

It should also be noted the world of assistive technologies and equipment can help us maintain habits and routines in new and different ways. I continue to be amazed at the variety of equipment available to patients and caregivers to help them renegotiate their daily routines.

At first, many patients resist use of such support, as they unfortunately forget strategies noted earlier in this book. For example, they may say "a hearing aid *is* a sign of old age" (word is not the thing) or "people who are disabled use wheelchairs" (map is not the territory).

But once patients can get past those semantic restraints, they can learn new ways to *modify* their daily routines with the help of their brains and neurological systems.

Shaking our habits is not easy. That's why we have clichés such as "you can't teach an old dog new tricks" and "old habits die hard." But our brains are quite capable of helping us imagine and learn new ways to negotiate our lives if we give them the opportunity to do so.

The following questions can help get that process started:

- Am I judging my habits as being "right" or "wrong?" If so, is that judgment correct?
- Is the way I do something the only possible way? What other behaviors might I adopt?
- What is preventing me from behaving in new ways? What will I gain from the new way? What will I lose?

Bumping into ourselves

Kenneth G. Johnson, a renowned General Semantics researcher and practitioner, suggested it is impossible for humans to talk about an objective world that is "out there." Instead, we can only talk about the "world to me" and the "world to you."

The nervous system, in both its glory and limitations, filters the experiences we attempt to describe and discuss with others. We cannot help ourselves when we speak to others, because we can only be ourselves.

Our role as mapmakers is an incredible one. Of course we want to create maps for ourselves that are accurate and will get us where we want to go. But rightly or wrongly so, we must remember any map we create is necessarily shaped by our assumptions, our habits, our levels of attention, and our interests and values.

Our goal then as mapmakers is to constantly survey the territories we encounter — comfortable or not — and use the biological tools we possess to revise those maps as needed.

We need to become explorers to some extent as we

enter new territories, especially those related to illness, and become aware of ourselves as we piece together new maps. By doing so, we can hopefully reduce the perils of the adventure and revel in new meanings we find along the way.

Chapter 5

The Great Return

"Nostalgia is a file that removes the rough edges from the good old days."
--Doug Larson, past editor of the <u>Green Bay Press-Gazette</u>

"When patterns are broken, new worlds emerge."
--Tuli Kupferberg

After absorbing the initial shock of a scary diagnosis, many patients quickly shift to wondering whether their lives could ever be the same again. Is there any hope of returning to what I once knew? Will I be able to pick up where I left off? Will I recover? Remission?

Certainly one does not want to remain sick. We want to recover as quickly as possible or at least hear reports suggesting our illness has waned. But when our conversations become solely focused upon future recovery or remission, we limit opportunities for other meaningful dialogue related to our lived experiences along the way. By focusing only on the *goal* of recovery, we miss out on conversations that could also help us find meaning in the *process* of recovery.

Here's a scenario to explain what I mean:

Imagine you and your partner had been saving money for years to purchase your dream home on the other side of town. According to your best-laid plan, you would be closing the deal within the next few months.

Then suddenly your neighborhood experiences a flooding in which both your own home and the one you had hoped to purchase were quite damaged. In fact, the new home suffered greater damage than your own. So you and your partner decide to use the money you had saved to buy the new home to instead repair your own home.

Much discussion and debate between you and your partner happens over time, but gradually you both realize you can make changes to your existing home that mirror the benefits you appreciated in the new home.

For example, you find a similar tile for the bathroom (although not exactly the same), you add a new pantry off the kitchen similar to that which you saw in the new home (although not exactly the same), you install fancy recessed lighting in the front room (although not exactly the same as in the new home), etc.

Making all those decisions definitely proved to be a bit

stressful, and not everything worked out easily. But when all was said and done, you and your partner decided to stay in your original home because all your careful discussions allowed you to essentially recreate the amenities you sought in the other house.

Would this end result suggest you had *recovered* your original loss? Perhaps, especially since your home is now almost identical to the new home you had planned to purchase.

But your renovations might also have actually resulted in a better home than you would have originally purchased because you took the time to adjust and accommodate throughout the renovation process. One might even suggest you went beyond recovery by virtue of the conversations you had during the *process* of recovery.

Now what if you and your partner had instead chosen to stick to the original plan—to wait for the owner to make all necessary repairs and then close the deal? Assuming that happened, would you have *recovered* your loss?

Likely so, but it's also very possible that the waiting period you experienced was more stressful than the other scenario because you were not an active participant in the recovery process. Instead, you were forced to wonder when, if, or how the owner would make changes. What if the owner short-changed the repairs? What if the price changed as a result of the renovations? How would you be able to afford repairs on your own home while waiting to purchase the other home? When would the home be ready? Etc.

By solely focusing on recovering that which you lost (your *goal* of recovery), your discussions were dominated by difficult questions and uncertainty because the owner was in charge of the recovery *process*.

Eventually things could work out and you would move into your dream home, but your sole focus upon the goal of recovery likely resulted in a more stressful experience and did not lead to the level of learning and self-discovery as did the process of recovery featured in the other strategy.

Two key points to take away from these scenarios:

(1) Something that is "recovered" may not be identical to that which was lost, and that's okay;

(2) The *process* of recovery often features more meaningful learning opportunities than the recovery itself.

Patients and caregivers need to learn how to blend discussions about recovery as a goal with discussions about recovery as a process. And for that to happen, conversations should feature discussion of *living with* or *adapting to* an illness as much, if not more so, as discussion about returning to a particular state of being before a diagnosis.

The balance of this chapter offers strategies for pursuing healthy conversations related to recovery and remission.

Replace "re-" words with "a-" words

A key semantic challenge we face in illness settings rests with our reliance upon words that start with the prefix "re-" to describe our preferred futures. After receiving a daunting diagnosis, we tend to cling to words like "recovery," "remission," and "return."

The challenge we face in using such "re-" words is they do not accurately describe lived experiences. The Latin prefix "re" means "back" or "backward" which is biologically not possible in either sickness or health. We cannot get younger, we cannot relive direct experiences, we cannot go back to what once was. Our bodies and emotions can only change over time in a forward motion, meaning we can never *fully* RE-cover or RE-gain the past.

Going back to my earlier scenario of the homes damaged by flood waters, neither of the renovated homes could be considered to be the same as the originals. Even if they appear to be identical, the fact is modifications have been made that prevent them from being fully identical.

The same can be said for recovery or remission in illness settings. Even though our bodies may no longer show signs of the presence of illness, that finding does not mean our bodies are the same as they were before the diagnosis. Accommodating and adjusting to the effects of an illness necessarily changes us not only in a physical sense, but also in an emotional and spiritual sense. We know things about ourselves and life that we would never have known otherwise by virtue of managing our illnesses. Therefore, a *full recovery* isn't possible, since we cannot revisit that moment of biological time and place prior to diagnosis.

Words like recovery and remission also ignore dimensions of change over time. They do not make room for talking about the diverse experiences we have as we strive toward recovery; they do not allow for discussion of our journey as an ongoing process.

Instead, each word suggests an "end state" of being that stands as a single, supposedly definable event that everyone watches for on the horizon. As patients and caregivers desperately scan the horizon for that event to happen, their conversations can become more stressful. Like the homeowners in the second example above, their conversations mostly feature questions of when, if, and how that come from a lack of personal engagement or control. Their sole use of "re-" words prevents them from talking about the moment-to-moment changes within the illness experience itself that could have meaning in and of themselves.

A good first step for talking about the present and future in illness conversations is to avoid using words with the prefix "re-" and instead use words that have an "a-" prefix. The table below shows some possibilities.

Avoid "re-" words:	**Use "a-" words:**
Recovery	Adapt
Remission	Accommodate
Regain	Adjust
Recoup	Assess
Rehabilitate	Acknowledge
Reclaim	Accustom
Redeem	Acquire

Note how the "a-" words are all verbs that point to a behavior that *largely moves forward*. Furthermore, "a-" words allow for the reporting of events *within and beyond* the illness experience. Patients and caregivers can use these words over time to discuss how they are responding to their illness experience on multiple levels (physically, emotionally, mentally, and spiritually) and create multiple meanings within those ongoing experiences.

Now look at the "re-" word list. It, too, features several verbs, but their direction of movement is constantly *backward*. The "re-" words don't allow for any sort of attention to the *present* illness experience, and as a result, they can't support conversations about the actual lived experiences. Because "re-" words inordinately focus people's attention on a past state that is not achievable, any possible meanings that could be found in living with the illness in its present form are lost to all involved in the conversation.

Below are several examples of statements that use "a-" words. Note that while they are all not necessarily positive sentiments, each statement acknowledges the passage of time and creates room for the possibility of changes in the future, including recovery.

"I've adapted my schedule to take into account I tend to have the most energy right after lunch."

"It's hard adjusting to the fact I cannot continue to work during my chemo regimen, but I'm trying to give myself permission to say 'I can't do that today.'"

"We have accommodated his inability to climb stairs by moving our bedroom to the first floor. It's definitely awkward, but it has helped a bit."

"I will assess my ability to continue participating in our book club as my treatments progress."

Each of the above statements leaves room for things to be different at a future date, depending upon how the illness progresses. And talking about the diverse ways in which adaptation and adjustment happen allows patients and caregivers to more accurately describe their progress or setbacks in a nonjudgmental fashion. None of the above statements suggest a right or wrong toward the accommodation being made. They simply describe one's behaviors and feelings at a particular time in the illness experience.

"A-" words also leave room for patients and caregivers to talk about what has been gained, along with what has been lost, while one adjusts and adapts to the illness. Knowing one cannot recreate or relive the past, "a-" words give permission to patients and doctors to accept what is and is not possible over time. No single winning event (i.e., recovery or remission) stands as the sole indicator of living well within an illness experience. Instead, even the slightest improvements can be held up as moments of success, which, in turn, helps reduce frustration and fear.

> "I can't pick up my grand-daughter, but we have adjusted by her climbing onto my lap when I'm in my wheelchair and we wheel around together. She thinks it's fun."

> "I'm so grateful to be able to be able to still visit the summer house. The kids created a small ramp to the beach that accommodates my walker. Roasted hot dogs have never tasted so good before!"

> "He was able to say what he wanted for breakfast out loud this morning, and I understood every word. We accommodate by using pencil and paper more often than not, but he seems really proud when he can say words here and there. It's so wonderful to hear his voice."

Finally, these "a-" words create opportunities to not only consider physical changes over time, but also mental, emotional, and spiritual changes. Reflective conversations allow patients and caregivers to talk about their new knowledge, feelings, and beliefs related to their having adapted and adjusted to the demands of their illness.

> "I'm proud of the way I've adjusted to my illness. People tell me how good I look, and I choose to believe them. Wearing wigs is a pain, but it helps me feel more normal."

> "At first I felt badly about not being able to take my kids to the pool. But we're all realizing it was the time together that was important…not the pool…so we've found new activities to do together that can accommodate my changing energy levels."

"It's been a big change having my husband needing me to do so many things for him, including helping in the bathroom. But we're adjusting, and I think we're finding what love for another person is really all about."

Even if a patient does seemingly recover from an illness or experiences remission, past use of "a-" words can help prepare for that life change, as well. When we purposely talk about changes we experience over time and reflect on the meanings those changes hold for us, our overall world views change too. We can never be the same person we were before the diagnosis *because* of the adjustments and accommodations we were forced to make along the way.

Our notions of priority, the importance of family and friends, and likely notions of life and death itself will all necessarily be impacted by the ways in which we talked about adjusting to our illness in the past. That's a good thing, as it demonstrates our ability to find positive meaning in seemingly negative life events.

Thoughts on "remission"

The word "remission" is just as stressful as "recovery" in that it describes an end-state of being without guarantee. The word creates an ongoing sense of imbalance that blends relief from news that no evidence of illness currently resides in one's body with anxiety that stems from constant anticipation of an illness's return. Just like waiting for the singular event of "recovery," the patient in remission anxiously watches for the singular event of the return of her illness.

Replacing the word remission with some form of the word "assessment" can help reduce the anxiety and fear of the situation. "Assessment" still finds a patient remaining watchful on an ongoing basis, but the assessment can relate to the *whole* self versus just that which is physically affected by illness.

For example, instead of focusing on the single question "will the cancer return?" a patient can instead ask multiple questions like "What am I physically feeling today?" "How is that similar or different from the past?" "Where am I mentally today?" "Where am I emotionally today?"

Assessment allows one to describe what one thinks and feels on multiple levels at a particular point in time, which can be more hopeful, and frankly more productive, than wondering whether or not an illness lurks within one's body.

Assessment also relates to using language that compares/contrasts experiences over time. By time-stamping observations, a patient can recognize inferences and judgments for what they are.

> "I remember when I had a pain in my left side last month, I was certain my cancer had returned. But after visiting the doctor, we decided it was likely stomach gas. This pain feels similar, so I'm going to keep an eye on it and assess how I feel in a day or two."

> "I guess I'll always wonder about getting sick again, but I feel great today, so I'm not going to assume something will go wrong anytime soon."

No matter what words a patient uses, a constant level of fear or anxiety exists for a patient in remission, especially during the first months and years of the experience. The patient is necessarily changed in that she now realizes good health is not a given, and because of her illness experience, she likely has lost a bit of confidence in her body and her future. In light of that, it will be important for her to make appropriate language choices when she talks about her fears about her illness's return.

A good first step is to avoid using single-word descriptors to talk about one's past life, as they can never be fully accurate. For example, saying one's life before cancer was "worry-free" and "safe" is not a true statement. Yes, there was an absence of a cancer diagnosis, but to suggest everything was fine and rosy is an untruth. Certainly there were negative events one faced during that time. Comparing or contrasting one's current or future life to a romanticized past is both inaccurate and unhelpful.

Patients and caregivers also need to avoid labeling one's life prior to an illness as somehow better or more stable and complete. The fact is we are always susceptible to illness or injury. We were then, we are now, and we will be in the future. So using language that suggests one's current and future lives are somehow inferior to that which was led before an illness experience is both inaccurate and unhelpful.

The following sets of statements demonstrate the difference between speaking about the past in untrue and aggrandized ways and simply making a comment based on observed behaviors.

Negative comment related to the past

"When I was my old self, I could jog 5 miles a day."

> ("Old self" is a high-level abstraction…what exactly is one's "old self?" The statement also suggests a level of superiority for the "old self" that diminishes the current self.)

Reframed comment related to the present and future

"I was able to run two miles today. I feel like I'm gaining more endurance."

> (The first statement solely focuses upon an observable behavior. The second statement is a judgment, however, it serves to offer encouragement to the speaker.)

Negative comment related to the past

"Before I got sick, I could wear low-cut tops and a two-piece swimsuit."

> ("Before I got sick" could go all the way back to birth! Thus, the statement is inaccurate. The statement also imposes judgment about one's ability to wear certain types of clothing. It implies not wearing those types of garments makes one less worthwhile.)

Positive comment related to the present and future

"I found a shirt that is a beautiful blue, and the v-neck is cut just right."

> (While the statement is judgmental in its word choices, all judgment relates to the clothing item versus the person wearing the item.)

Negative comment related to the past

"Life was so much easier before I had cancer."

> (The phrase "so much easier" is ambiguous and

impossible to prove. It suggests one had little or no problems in life which likely is untrue. "Life" is also a squishy word. Does the speaker mean physical life? Social life? Spiritual life?)

Positive comment related to the present and future

"I have had to learn how to do some things differently based on my changing energy levels."

(No judgment is made as to whether or not the changes a patient has made are worse or better than life experiences in the past. The statement simply explains the patient's current experience.)

During times of seeming remission, it is also important to separate facts, inferences, and judgments when talking about one's experiences. Remember, facts can be proven, inferences are statements made about the unknown based on the known, and judgments place value or worth on statements.

People living after serious illness are especially prone to inferences and judgments. They easily interpret *current* experiences (unknown) based upon their *past* experiences (known). This can be stressful when one forgets the map is not the territory, and what one experienced during the actual illness experience cannot be fully identical to current or future living experiences.

Inaccurate statement

"This is the exact same pain I had before I was diagnosed. I think the cancer has returned."

(Using the words "exact same pain" is inaccurate, as instances of pain are biologically separate and unique. Also, the phrase "before I was diagnosed" is a broad timeframe. And the last statement is not only an inference, but also a fairly broad assumption. Memory also compromises the validity of the statement. How accurately can someone really remember a specific pain?)

More accurate statement

"I've got a pain on my left side. If it doesn't go away within the next 48 hours, I'll call my doctor."

("I've got a pain on my left side" is factual. No inferences or assumptions are made about the source/meaning of the pain. The speaker instead focuses on possible action steps in response to the pain.)

Inaccurate statement

"My friend's husband's cancer returned, and he had a similar diagnosis to mine. This is bad."

(The first sentence does feature facts in terms of being able to measure the presence of cancer and similar diagnoses. However, the speaker forgets actual cancer experiences are never the same, and draws a judgment based on an inference that he, too, will see his cancer return. He also makes the assumption based on memory and fear from past experiences that the friend's husband will also have a bad experience.)

Accurate statement

> "My friend's husband's cancer has returned, according to what his doctors told him."
>
> (The statement features information that can ultimately be proven, but leaves room for possible error in judgment by the doctors. The speaker also draws no inferences from the other person's experience related to his own future health.)

In summary, facing a diagnosis of a life-limiting illness finds us wanting to bend life's rules, especially in terms of being able to go back to the lives we led before receiving news of our illness. Our efforts to remain invincible overshadow what we have always known about time and space. We forget we have always faced the possibility of illness or other negative life event.

To better prepare ourselves for whatever may happen in the future, we need to talk in ways that accurately describe the past, present, and future. That means our conversations need to lean toward talk of recovery or remission as a *process* while keeping the goals for both in sight, and we need to avoid inaccurate inferences and judgments. The following questions can help you better say it like it is between the time of diagnosis and hopeful recovery:

- Do the words I use reflect the forward-moving nature of life? Do my word choices allow for change over time?
- Am I describing my ongoing experiences as I strive for recovery or remission (versus focusing solely on the singular events of recovery/remission)?

- What adjustments or accommodations have I made as I strive for recovery or remission? What have I learned as a result?
- Am I avoiding drawing comparisons between life before and after my diagnosis?
- How are my memories influencing what I am saying? Is that appropriate?

Chapter 6

Pain: What Words Can Never Fully Express

"Whatever pain achieves, it achieves in part through its unsharability, and it ensures this unsharability through its resistance to language."
-- Elaine Scarry[1]

(Patient rises up out of chair.)
"Oh...my...stars! This hurts!"
"What hurts? Your back? Your legs?"
"I don't know what it is...but it feels like it's in this area!"(Points to right hip)
"Like a cramp?"
"Like a knot tied too tight...it's just gripping me. Hard to move my leg."
"Have you felt it before? Why do you think it's happening?"
"I don't know!"
"I'm just trying to understand..."
"Ouch!"
"Tell me what you need!"
"I don't know. It just hurts like heck!" (Drops back into chair)

The small, four-letter word "pain" is possibly the most profound example of the gap that exists between what we *say* and what we *mean* in illness settings. There literally are no words that can fully describe the varieties of pain an illness causes, and ironically, our frustrated attempts to explain what we feel can actually exacerbate the pain we are trying to describe. Trying to report what hurts, how it hurts, when it hurts, etc. exposes the limitations of the symbolic nature of language. The words we use simply cannot fully represent that which we experience.

Our first challenge in describing pain rests with the fact that our word choices for aches and pains tend to be limited to those used within routine illness experiences: a "stuffy" nose, a "throbbing" headache, an "upset" stomach.

Because we share many of those illness experiences and use similar descriptive terms, we basically understand one another in such conversations. We respond as if we *know* what the other person is saying.

But when faced with the unique pains of a serious illness and/or its treatment, that repertoire of descriptors quickly becomes inadequate. Patients experience pains like never before, and their circles of support are even less likely to have experienced such pain. Even if a family member or friend has shared a similar diagnosis, pain is unique to everyone. Therefore, the words patients relied upon earlier simply don't work, and unfortunately, continued use of those misplaced words and phrases can lead to misunderstanding and inappropriate responses.

A second challenge related to talking about pain is that caregivers are forced to operate solely on that which the patient describes. Caregivers cannot actually experience the patient's pain, so they find themselves working hard to ask for and listen to patient's pain reports.

But those reports are necessarily interpreted through (and limited by) the *caregiver's* personal experiences and their prior use of the words and phrases featured in those reports. Remember the "personal spins" we all put on words and phrases? The same challenge rests with discussions of pain.

For example, if a patient reports a "pounding" pain, the caregiver may think that pain is akin to a "pounding" headache the caregiver experienced before. Armed with that personal interpretation, the caregiver will then talk to and behave toward the patient in light of the caregiver's experience. If the caregiver did not think the pounding headaches she experienced were all that terrible, the caregiver might view the patient's response to the pounding pain as overly dramatic. But if the caregiver's experiences of pounding headaches were quite awful, then the caregiver would be inclined to offer more comfort and sympathy to the patient.

This phenomenon also points to the dangers of inferences and judgments that come to life in conversations about pain. One's threshold for pain becomes exposed through direct reports of patients and then later becomes ripe for judgment by both patients and their circles of support.

Patients worry that talking about pain means they are not as strong as they think they should be in the wake of painful experiences. Patients are also concerned about being labeled as "complaining" if they talk about pain, or worse, simply not being believed by others when attempting to describe unseen pain.

In turn, caregivers worry their inability to physically relieve the pain will be exposed through pain discussions, so they may also avoid such conversations. When caregivers do pursue discussions of pain, they then worry about saying the wrong thing that might make the pain experience worse. They walk on eggshells hoping they don't say something that will further frustrate the patient.

Inferences can be equally dangerous. Inferences come to life when a patient does *not* communicate pain experiences, whether out of fear of facing judgment or because the pain simply ebbs and flows. Without direct communication about pain from patients, caregivers often infer a patient's pain experiences based upon observing the patient's behaviors (i.e., grimacing, wincing, etc.) and/or based upon that which the caregiver believes the patient should expect to feel. When the caregiver's inferences do not represent the patient's actual lived experience, both the patient and caregiver risk offering inappropriate responses to each other.

Finally, caregivers find conversations about pain to be especially difficult because the caregiver feels "helpless" in light of those discussions. Caregivers often believe the patient is reporting pain in hopes that the caregiver can do something about those pains—that the caregiver can fix the problem. In many instances, caregivers cannot provide the physical relief they believe patients are seeking, which leaves caregivers feeling inadequate and failing.

Despite all these pain discussion challenges, most research suggests it is better to pursue "painful conversations" versus not talking at all. Pain can be debilitating when left untreated, not only for patients, but also for their circles of support. When pain is discussed openly and carefully, it can be better managed which, in turn, creates a more comfortable way of living for everyone.

Talking about pain can also help reduce a patient's sense of isolation and loneliness. The fact no one else can share another's pain can make patients feel separated from family and friends. Patients feel no one can possibly understand what they are experiencing, so they draw further and further into themselves.

But research shows talking about pain with loved ones can actually make patients feel more connected and understood. While pain may persist, patients are better able to cope, because they have the emotional support they need to do so.

Caregivers can also reduce their feelings of helplessness by engaging in meaningful conversations about pain. By providing the emotional and spiritual support patients need, caregivers can feel as if they are doing something related to the pain, and relationships among everyone are strengthened as a result.

Following are more specific strategies for talking about pain that relate to the GS principles of the word is not the thing, the map is not the territory, and the map reflects the mapmaker.

Learning how to describe pain

The medical field has long been challenged to create ways for patients to accurately describe their pains so effective treatments can be pursued. Research efforts in that regard have resulted in a variety of word lists and sets of visual images that patients can refer to when reporting pain to their doctors.

Two of those tools can also be helpful when talking about pain at home. First, the "McGill Pain Questionnaire"[2] features a fairly expansive word listing that can help patients better pinpoint the sensations they are feeling. Second, the Missoula Demonstration Project's "pain assessment scale and checklist" features a more limited word list, but it provides good discussion questions for talking about pain experiences with family and friends.

The McGill Pain Questionnaire was developed by Dr. Ronald Melzack and Dr. Warren S. Torgerson in 1975 to help the medical community move from focusing solely on the intensity of pain to more so exploring the varied dimensions and characteristics of pain. By only talking about pain intensity, the researchers suggested doctors limited themselves in prescribing effective pain treatment. The practice ignored other important sensations of a pain experience, and as a result, limited the doctor's creative responses to fully managing the pain.

The researchers developed a pain vocabulary—a word listing—based upon words they heard patients use to describe the multiple sensations they experienced in single pain episodes. The word listing was then organized into four categories: (1) sensory words, (2) emotional or feeling words, (3) judgment or evaluation words, and (4) miscellaneous words.

The sensory word group features words that largely relate to the timing and duration of pain. For example, a "flickering" or "pulsing" pain suggests a different timeframe from a "boring" or "drilling" type of pain. Sensory words also point to types of pressure a patient might feel, such as "crushing" or "pulling," as well as thermal or heat-related words related to the pain, such as "burning" or "stinging."

The emotional or feeling words list revolves around tensions, fears, and/or "punishment" a patient feels from the pain. Note how the list includes several words that almost personify pain by describing it as "cruel" or "nagging." The pain is described as an outside agent, to some extent.

McGill Pain Questionnaire

Sensory Words	Sensory Words	Emotional/Feeling Words	Miscellaneous
Flickering	Cramping	Tiring	Spreading
Quivering	Crushing	Exhausting	Radiating
Pulsing	Tugging	Sickening	Penetrating
Throbbing	Pulling	Suffocating	Piercing
Beating	Wrenching	Fearful	Tight
Pounding	Hot	Frightful	Numb
Jumping	Burning	Terrifying	Drawing
Flashing	Scalding	Punishing	Squeezing
Shooting	Searing	Grueling	Tearing
Pricking	Tingling	Cruel	Cool
Boring	Itchy	Vicious	Cold
Drilling	Smarting	Killing	Freezing
Stabbing	Stinging	Wretched	Nagging
Lancinating	Dull	Blinding	Nauseating
Sharp	Sore		Agonizing
Cutting	Hurting	**Judgment Words**	Dreadful
Lacerating	Aching	Annoying	Torturing
Pinching	Heavy	Troublesome	
Pressing	Tender	Miserable	
Gnawing	Taut	Intense	
Cramping	Rasping	Unbearable	
Crushing	Splitting		

The third group of judgment/evaluative words largely speaks to the intensity of the pain. Words range from a mild to severe reaction to the pain. In addition to the word choices, the list also has a grouping that represents a five-point scale of intensity ranging from "no pain" to "excruciating" pain (not featured on the chart).

Finally, the list of miscellaneous words seems to be a cross-section of sensory and affective words. I am not sure why these words were given their own category, but still, they are useful to consider.

Doctors who use the questionnaire ask patients to identify any and all words that relate to the patient's pain experience across all three categories. Doctors are then better able to diagnose and treat pain symptoms based upon the more comprehensive set of descriptors patients provide. While use of this tool at home will not enable anyone to necessarily diagnose or prescribe medication, the clarification of pain sensations can lead to more productive conversations about pain.

McGill's word lists are hardly exhaustive, but they do offer a good variety of descriptors we can use to help explain the seemingly unexplainable. Statements such as "I feel *lousy*" or "I feel *awful*" provide limited information, and as a result, solutions to the pain will also necessarily be limited. For patients and caregivers tracking pain experiences, use of generic descriptors also makes it difficult to authentically track changes in pain over time. Many discrete feelings and sensations could fall under the labels of "lousy" or "awful," each requiring its own specific response for relief.

Patients and caregivers need to work harder at avoiding bland reports of pain, just as they do when they visit the doctor. They need to develop an arsenal of words that can more specifically explain the pain experience.

Patients should also feel free to add other words to McGill's lists that work better for them or come closer to their shared meaning. The more word choices possible, the greater the likelihood of increased understanding.

Lastly, patients and caregivers are encouraged to combine words across the four categories as needed, recognizing that pain is not a single-level experience. Pain comes to life in virtually all aspects featured on the questionnaire.

Another resource comes from the Missoula Demonstration Project which combines a "Pain Assessment

Scale" with a "Pain Checklist" to help patients describe their pain experiences (see figure on next page). This tool was created in 1998 as part of a community education effort that largely targeted seniors living in Missoula, Montana.

The Project wanted community members to understand that pain management is an essential part of health care and that patients' abilities to effectively communicate their pain experiences was critical. To that end, the Project developed a handy bookmark-sized tool featuring an intensity scale and pain checklist for users to refer to when reporting pain to their doctors.

The "intensity" scale ranges from 1 to 10 and features both facial expressions and verbal descriptors. The facial illustrations come from the "Wong-Baker Faces Pain Rating Scale" that was developed to help children and/or those who did not speak English an easier way to describe their pain by pointing to the matching facial expression.

The Project used that facial scale and then added descriptions in terms of one's ability to participate in daily activities. The descriptor choices are not nearly as expansive as McGill's word lists, but they do offer an interesting and simple way to view the impact of pain.

The checklist then builds on intensity measures by posing key discussion questions to better pinpoint the pain experience. Specifically, questions are asked about the location of the pain, the onset and duration of the pain, and the "quality" of the pain. The combination of questions can provide a good foundation for fostering more meaningful discussions of the pain experience for both the patient and caregiver. The questions are also helpful in tracking changes in pain experiences over time.

Again, the overarching goal of using tools such as word lists and discussion questions is to help patients and their circles of support talk about pain in more accurate ways. Doing so could result in more effective pain management while also reducing frustrations that can

actually increase patients' pain and caregivers' anxieties.

Missoula Demonstration Project's Pain Scale and Checklist

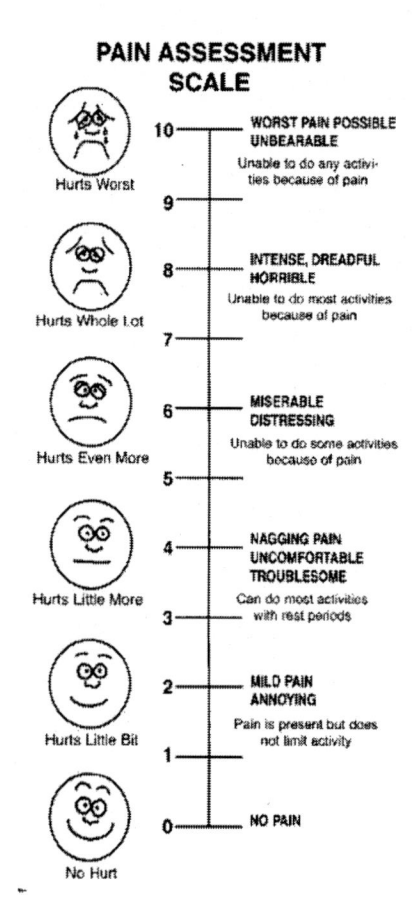

PAIN ASSESSMENT SCALE

Hurts Worst	10	WORST PAIN POSSIBLE UNBEARABLE — Unable to do any activities because of pain
Hurts Whole Lot	8	INTENSE, DREADFUL HORRIBLE — Unable to do most activities because of pain
Hurts Even More	6	MISERABLE DISTRESSING — Unable to do some activities because of pain
Hurts Little More	4	NAGGING PAIN UNCOMFORTABLE TROUBLESOME — Can do most activities with rest periods
Hurts Little Bit	2	MILD PAIN ANNOYING — Pain is present but does not limit activity
No Hurt	0	NO PAIN

PAIN CHECKLIST

LOCATION: Where does it hurt?

INTENSITY How bad is the pain? (*Rate* your pain using pain scale on other side)

- Now ———— (0-10)
- Worst ———— (0-10)
- Best ———— (0-10)
- Acceptable ———— (0-10)

ONSET: When did the pain start?

DURATION: How long have you had this pain?

QUALITY: Is it constant or on-and-off? Dull or sharp? Burning or pressure?

What makes it worse?

What makes it better?

Does it affect your usual daily routine? Sleep? Concentration? Mood?

This pain scale was developed by the Missoula Demonstration Project Pain as the Fifth Vital Sign Task Force, a project supported by the Mayday Fund, 1999.

320 E. Main
Missoula, MT 59802
(406) 728-1613

Wong-Baker FACES Pain Rating Scale from Wong DL, Hockenberry-Eaton M, Wilson D, Winkelstein ML, Ahmann E, DiVito-Thomas PA: *Whaley and Wong's Nursing Care of Infants and Children*, ed 6, St. Louis, 1999, Mosby, p.1153. Copyrighted by Mosby Year Book, Inc. Reprinted by Permission.

Reprint permission granted by Duke Institute on Care at the End of Life

Use of metaphors and analogies

Some research suggests using metaphors or analogies can also help patients describe their pain. Metaphors and analogies are essentially word pictures—often figurative in fashion—that try to make complicated ideas easier to understand. For example, a patient might say "I feel like I've been run over by a truck" or "This pain is worse than childbirth."

While metaphors and analogies can be helpful, they only work if both the person talking and the person listening have experience with the similarities being discussed. Granted, most of us have not been run over by a truck, but because the phrase is fairly familiar in our society, for some of us, the analogy will work.

But when it comes to men trying to understand childbirth analogies, those could be a stretch. Certainly men have overheard women sharing labor room war stories, but men simply are not emotionally, let alone physically, equipped to fully understand that type of pain experience.

The key then to using metaphors and analogies is to remember they are not foolproof, so be prepared to clarify as needed.

Still, there are some metaphors or analogies that should be avoided completely, especially those that suggest one's body is a mean-spirited entity that has somehow "betrayed" itself. I have often heard patients remark they feel their bodies have "turned on them" or otherwise "let them down" when their pain is pervasive and ongoing.

As noted earlier in this book, our bodies are complex groupings of systems, all of which face the possibility of illness or disability at some point in time. To suggest then that the body is doing something to itself—that it has control over its functions—is an inaccurate statement.

There has never been a time in our lives when we were totally immune from illness, and furthermore, the body is not capable of fulfilling a role that requires a sense of "trust." Our bodies are not independent of ourselves, and if anything, a more accurate metaphor would be to suggest our bodies need to *trust us* to help sustain them.

Outrage or disappointment in our body's aches and pains are certainly natural responses, especially when we believe we have done our best to stay healthy and "normal." All that exercising and eating right was supposed to protect us, yes? Possibly, but again, our bodies were, and continue to be, susceptible to injury and disease at all times.

We are also tempted to feel betrayed or angry at our bodies when we experience pains like never before. While most of us are used to having a terrible headache now and then, a toothache, a strained muscle, etc., we rarely suggest our bodies have somehow betrayed us during those pain episodes. But pain that has never been experienced before is both surprising and frightening, and it can definitely make us wonder how our bodies can produce such incredible discomfort. Still, we need to check the expectations we hold for our bodies, and our conversations must focus more on ways to manage pain versus laying blame upon our bodies and ourselves.

Negative Metaphors/Analogies

"My back just won't leave me alone!"
"My knees just don't want to work right anymore."
"His stomach has just turned on him."

Reframed Statements

"I have a gnawing ache in my lower back that just

won't let go. I cannot seem to get comfortable no matter how I sit."
"My knees have been swollen for the past two weeks."
"He cannot keep food down lately."

Notice how the first set of metaphors comes across as fact, and as a result, they stifle any real follow-up conversation. The body has turned on itself, period.

But the second set of reframed statements leaves room for discussion of possible solutions to the problem. "What can we do about that?" seems to be a natural follow-up statement that opens the door for further discussing ways to manage pain.

We must work to only describe that which we sense or observe, and any attempts to infer "cause" should only relate to discussions about managing the pain.

The danger of negative "pain maps"

My dad never asked for novacaine or gas when he had dental work. He said the "fuzzy mouth feeling" that lingered after a filling was hardly worth the short amount of pain he experienced in the process. I was stunned he would choose to suffer through such pain, and admittedly, wondered if I was somehow weak in not doing the same (though he never suggested that was the case).

My dad's response to pain was likely influenced somewhat by his father who also tended to slough off pain. My grandfather's pat responses to pain were to "buck up" and "shake it off." Crying got you nowhere. Scrape your knee on the sidewalk? Brush it off. Chip your tooth while wrestling with your brother? Consider it a badge of honor.

On the other hand, the women in my family were much more solicitous. My maternal grandmother would pull you onto her lap and yell at the sidewalk for scraping your knee. Or she'd offer home-baked cookies to help ease the sting of a paper cut.

As children, we are taught how to respond to and talk about pain by the significant adults in our lives. Those pain "maps" can become deeply embedded in our psyches, especially if we have multiple experiences with pain and illness while growing up.

When those maps continue to work for us, little, if any, stress is experienced. But when those maps do not match our lived experiences, especially related to pain, we can become distressed and anxious.

In my hospice volunteer experiences, I see a real resistance to use of pain medication among seniors. Patients have told me they felt taking pain meds demonstrated a weakness on their part, usually remarking "there are plenty of people in this world who are worse off." Many seniors also report serious fears about becoming addicted (like the addicts they saw on TV and heard about in the news).

Most of the patients I'm describing grew up during hard economic times, often on farms, where injury and illness needed to be dealt with as quickly as possible to keep chores on track. Taking time off could literally threaten a family's livelihood, and talking about pain didn't help get the cows milked or put dinner on the table. Sadly, many of those patients' verbal maps about how pain should be managed caused them to forfeit any benefits they could receive from today's pain regimens, and for some, their quality of life was severely diminished.

Many of those patients' family members were also armed with similar maps which made them resist asking nurses and doctors about pain management possibilities for their parents. First, if their parent was unwilling to ask for help, the children were hardly going to challenge their parent. But adult children were also concerned pain medications would somehow make their parents behave differently or be "less present." They were desperate to instead keep things as "normal" as possible.

Despite increased advertising for drugs designed to ease both physical and emotional pain, many of us still operate under more long-standing verbal maps that suggest we should avoid discussions of pain and its management:

> No pain, no gain. *(said to be adapted from Benjamin Franklin's saying "there are no pains without gains" in the* Poor Richard's Almanac: The Way to Wealth*)*
> Pain is temporary; quitting lasts forever. *(Lance Armstrong)*
> Pain is inevitable; suffering is optional. *(Zen aphorism)*
> Pain and death are part of life. To reject them is to reject life itself. *(British psychologist Havelock Ellis)*
> Pain is weakness leaving the body. *(source unknown)*

Verbal maps like these are terribly hard to shake, but when it comes to managing pain, we must create more accurate maps for ourselves. We must especially learn to remove judgment from our maps of pain, particularly related to one's threshold for pain.

Simply put, a body in pain has trouble healing. Muscles tense, circulation and breathing are compromised—the body cannot relax and attempt to mend itself. That physical tension can also create *new* pains as the body attempts to adjust to existing pains. For example, one might favor using the right leg in response to pain in the left hip, only to later damage the right knee due to overuse!

Our emotions are also affected by pain, especially ongoing pain that can lead to depression and anxiety. Pain can consume our every waking thought, which prevents us from engaging with others, or worse, prevents our ability to maintain a sense of hope. Pain pulls at our soul just as much as it does our body.

A patient's pain also affects her circles of support at home or at work. Family and friends may need to take on new roles, routines may be disrupted at times, children can grow fearful seeing a parent in pain, etc. Stress levels increase for everyone when a patient is in pain.

The good news is these painful outcomes do not have to be the norm. Pain management research and techniques have lead to a wide array of options, many of which are drug-free. But to tap into that information, we must first create new verbal maps about pain—maps that remove judgment about how pain can or should be addressed and that grant us *permission* to talk about what we are feeling and to ask for help.

Here are examples of new verbal maps related to pain I would like to see come to life:

I cannot be wholly present to friends and family when I am in pain.

It is okay to use medications or other accommodations to reduce pain.

Pain is not something I have to endure.

Seeking pain treatment is not a sign of personal weakness. Pain hurts!

People have unique thresholds for pain, none of which are better or more correct.

The tipping point

A final key challenge to talking about pain relates to patients' concerns they will be viewed as "complainers." This is especially true for patients with chronic pain—pain that does not seem to get any better over time.

Society tends to disapprove of people who complain, especially if complaints are believed to be unfounded or if the subject of a complaint cannot be fixed. We believe if nothing can be done, adjust accordingly, move on, and quit with the complaining.

Complaining is linked with other frowned-upon behaviors, such as whining and nagging, and some people view complaining as a negative way to get attention from others.

Perhaps our society's distaste for complaining comes from our agricultural heritage. One cannot stop torrential rains or an early hard frost, so get out there and harvest what you can. Or perhaps our response to complaining grew as a matter of etiquette. Burdening others with our personal problems is quite impolite. Whatever the source of our disdain for complaining, it is safe to say the word "complain" rarely bears a positive connotation.

Complaining in illness settings seems to be no exception, but the process follows an interesting trajectory, especially for patients whose aches and pains do not go away.

When a patient is first diagnosed with a serious illness, family and friends rally around the patient and ask her questions like "how are you feeling?" and "are you in any pain?" Caregivers are especially eager to help, so they repeatedly pose such questions, sometimes hour to hour, wanting to gauge a patient's recovery.

In turn, the patient assumes those questions are her cue to report her pain openly and honestly. And as discussed earlier in this chapter, she finds relief in doing so, because she gains the emotional support she seeks.

But when an illness lingers and pains do not go away for an extended period of time, the patient's continued reports of pain can morph into what the patient and/or the caregiver view as *complaints*. When and how that change happens is unique to every relationship, but it is usually grounded on a patient's having to repeatedly say "I still have pain" and a caregiver's continued inability to do anything about it.

This tipping point—when pain reports turn into complaints—can severely hamper effective discussions about pain and result in hurt feelings, resentment, and despair.

The tipping point is largely founded upon frustrations both the patient and her circle of support experience when pain does not go away. After all the trips to the doctor, all the medications, all the accommodations at home, etc., nothing helps. Pain just seems to have become a part of life for all involved.

Unfortunately, once such an idea is internalized, everyone then succumbs to our society's belief that if the pain isn't going to go away, everyone should learn to live with it and quit complaining.

Ironically, research on complaining suggests that when all seems lost, the act of complaining can actually be beneficial. It allows people to vent and rage at what *is* seemingly out of their control, which in itself can be therapeutic. If done well, complaining can actually give people the emotional release they need while at the same time gaining the relational support they need from family and friends.

If not done well, complaining can tear relationships apart and cause anger, resentment, and a cadre of other negative feelings.

So how does one complain in a good way? And how does one respond to complaints in a positive fashion?

First, we must remember complaints are reports about the feelings we experience when faced with a negative event or situation. They are statements that tend to be emotionally-charged, especially if we feel we have been wronged or hurt somehow, but they can also feature rational evidence and be delivered in a calm manner.

Based on my description of a complaint, read through the list of phrases below and identify which phrases are "reports of pain" and which are "complaints:"

"I have a shooting pain in my left calf."

"My fingers just sting from the cracks and peeling skin."
"Everything I eat hurts the sore in my mouth."
"The nausea just never goes away."

I'm guessing you cannot sense a difference, at least upon first read. That's because you have no familiarity with the patient speaking or the context within which the statements are being made. You also cannot hear the tone of voice or see facial expressions as the statements are made, and you do not know how often a phrase has been repeated. All you have to work with are the words themselves, and therefore, at face value, they are simply reports of pain and discomfort.

Now, if I were to tell you a patient has been saying any of the above phrases every day for the past three weeks, does that turn that report into a complaint? What if the phrase is mentioned every waking hour? Does that turn the phrase into a complaint? What if the phrase is the only communication you hear from the speaker—she doesn't speak of anything else?

On the other hand, what if the speaker says a phrase in passing, perhaps only if asked directly how she is feeling? Or what if you only hear a phrase spoken when the patient is visibly in pain? Is she complaining?

While these contextual cues may not make your decision easier, I imagine you were more willing to consider the possibility the statements could be complaints in light of the frequency of those reports. And that room for possibility illustrates the phenomenon…and danger…of a tipping point in our conversations.

To help alleviate the possibility of reports turning into complaints, patients and caregivers should first focus solely on the words and phrases in the reports themselves and not add extra layers of meaning (inferences) to those reports (just as they did when the patient was initially diagnosed). They should not think about how many times each has heard or said similar reports, nor should they try to connect the particular report about pain to other reports about pain. Patients and caregivers must try to keep discussions about pain in the present tense as much as possible and work only with statements being made in that particular conversation.

Certainly it is not easy to ignore all the familiarities and contexts that have developed over time as pain has lingered. But if patients and caregivers can agree to set the past aside, patients will feel safer in speaking about their current pain experiences, and caregivers will only have to listen and respond to the patient's current state of being.

Another concept related to the complaint tipping point is we tend to complain in direct proportion to the reaction of our listeners. Complaining is a reciprocal process in which we complain (or not) to the extent people will listen (or not). If a listener seems receptive to hearing our complaint and becomes engaged in the conversation, we will likely continue to complain. On the other hand, if the listener seems to avoid such conversations, we will eventually stop complaining, as we are not gaining the emotional support we seek.

Our decision to complain also relates to the type of relationship we have with a listener. If we care deeply about maintaining our relationship with a listener—such as a spouse or parent—we will be much more careful in not positioning ourselves as complainers. The same can be said if we rely on someone to a large extent for help and support. We will curtail our complaining immediately if we sense the possibility of losing that support.

Trying to figure out that tipping point within an important relationship is both fearful and exhausting, so oftentimes, both the patient and caregiver choose not to complain (to not report pain) to remove any threat to their relationship. But as discussed earlier, not talking about pain can actually damage the relationship more than an open and honest discussion of pain.

So the second step to avoid surpassing the tipping point is to agree to listen to one another's complaints. Note I only use the word "listen." I do not suggest anything else must happen. Listeners do not have to agree with the complaint, they do not have to fully understand the complaint, and they do not have to do anything about the complaint. The listener's job is to simply listen and be present to that person.

Some have referred to this type of listening as offering a "compassionate presence" to those we care about. The greatest gift we can give to someone in pain is to listen with our ears, our minds, and our hearts. To do so, we must eliminate any distractions, maintain eye contact, ask for clarification if needed, and remind the patient how much we care about her.

Building on that need for compassionate presence, a final point to remember related to complaints is we are more than willing to complain even when we know the person listening cannot do anything about our problem. We are social beings, so we talk about what's wrong in hopes that family and friends can offer us emotional support. We want someone to be on our side and tell us we have a right to be dissatisfied. We want to know someone is listening to us and that we matter.

This, too, points to the potential benefits of reporting pain (complaining). Patients know their pain cannot be borne by others, but there is still something therapeutic in being able to share our frustrations, disappointments, anger, etc. with those whom we care about.

A final strategy for helping patients and caregivers avoid reaching the tipping point is to craft an agreement about how they will talk about pain that lingers over time. Not talking about pain is not an option, so a plan is needed that meets the needs of both the patient and the person listening. Following are key areas to consider for developing such an agreement:

- Try to identify good times to talk about pain in which both the speaker and listener can be fully present for one another. Maybe you want to regularly "check in" with each other for 15 minutes at the beginning and end of the day. Or maybe you agree to discuss pain when noticeable changes happen over time. Perhaps you agree not to talk about pain at family gatherings and meals or that you do not want to hold such conversations in the presence of young children. Maybe you decide to have the patient journal pain reports first, and then she can choose which to share at a later time. The key is to try to plan for conversations that can afford all the benefits of sharing in shorter periods of time.
- Try to use "I-statements" when issuing a complaint. The goal is to avoid blaming or pointing fingers at others. An I-Statement uses this formula: *I feel (insert feeling word) when (insert event/situation) because (insert description of impact on yourself)*. For example, a patient might say "I feel disappointed when I wake up each morning with the same sick stomach because it limits my ability to accomplish what I want to do each day." This complaint is not only specific, but it also keeps responsibility and possible action related to the complaint on the speaker's side of the fence. The listener is not implicated as being the cause of the problem or as needing to find a solution to the problem. The speaker essentially "owns" the complaint, and as a result, feelings of defensiveness or helplessness on the part of the listener are reduced, and the conversation can move forward.

- Agree that the listener has a choice in trying to respond to the complaint. Listeners should not feel obligated to fix things, and they should recognize they provide valid and important support just by being a compassionate listener.
- Agree that both patients and members of their circles of support all have the right to complain about how a patient's pain affects them. No one gets to complain more or less. Everyone's frustrations matter, and everyone can support one another.
- Agree to believe each other's reports, especially related to feelings of pain. Pain is not always directly observable by others — it exists at neurological levels that are not necessarily evident through visible wounds or other indicators. Pain reports should not be contested; they should be accepted as the speaker's personal truth.

In conclusion...

Talking about pain will rarely come easily for either the patient or the caregiver, but it is still important to try to broach the subject. It is true one cannot talk one's self out of physical pain, but healthy pain conversations can provide at least some level of emotional and spiritual support that can help ease suffering.

The following questions can help both patients and their circles of support modify or create verbal maps of pain that can pave the way for more effective conversations with others. Honest reflection on the questions will also hopefully lead to more effective pain management that leads to a better quality of life for all.

- How can I best describe the pain I feel? What words are most accurate?

- How is pain affecting my ability to live a comfortable life?
- What does the pain prevent me from doing?
- What do I know about pain management options? How could I learn more?
- What judgments do I make about my ability to deal with pain? How can I stop making those judgments?
- How are my loved ones' responses to pain influencing how I respond to pain? Are their responses appropriate for me?
- Aside from my doctor, who can I talk with about my pain? What would be the best way for us to pursue conversations about my pain?
- How does my pain experience affect those I care about? What would be the best way for us to pursue conversations about their experiences?

References

Chapter 1

[1] Johnson, K. G. (2004). *General semantics: An outline survey.* Fort Worth, TX: Institute of General Semantics.
[2] Korzybski, A. (1958). *Science and sanity: An introduction to non-aristotelian systems and general semantics.* (4th ed.) Brooklyn, NY: Institute of General Semantics.

Chapter 2

[1] Sternburg, J. (2003). *Phantom limb.* Lincoln, NB: Bison Books.
[2] Johnson, Wendell. (1956). *Your most enchanted listener.* New York: Harper.

Chapter 3

[1] Johnson, Wendell. (1956). *Your most enchanted listener.* New York: Harper.
[2] Frankl, V. E. (1984). *Man's search for meaning: An introduction to logotherapy.* Boston, MA: Beacon.
[3] Koestenbaum, P. (1976). *Is there an answer to death?* Englewood Cliffs, NJ: Prentice Hall.
[4] Estess, J. (2004). *Tales from the bed: A memoir.* New York, NY: Washington Square Press.
[5] Price, R. (2004). *A whole new life: An illness and a meaning.* New York, NY: Scribner.
[6] Sacks, O. (1984). *A leg to stand on.* New York, NY: Touchstone.
[7] O'Kelly, E. (2006). *Chasing daylight: How my forthcoming death transformed my life.* New York, NY: McGraw-Hill.

Chapter 6

[1] Scarry, E. (1985). *The body in pain: The making and unmaking of the world.* New York: Oxford University Press.
[2] Melzack, R. (2005, July). The McGill Pain Questionnaire: From description to measurement.
Anesthesiology, 103(1), 199-202.